HFC

Hydrogen Fuel Cell Cars

*The Next Generation
in Electric Cars*

BOB KAPHEIM

HFC
Hydrogen Fuel Cell Cars
The Next Generation in Electric Cars

© 2024, Bob Kapheim

Print ISBN: 979-8-35096-565-0

eBook ISBN: 979-8-35096-566-7

INTRODUCTION

B Y 2023, ALMOST ALL MANUFACTURERS HAD put on the market a full range of electric vehicles. There was a full availability of small, medium size and large SUVs as well as small, medium size and large sedans. The government was pushing hard to make the transition from fossil fuel to electric vehicles. In 2023, President Biden praised Hertz for their effort to include electric rental cars in their portfolio. In 2024 Hertz announced they were selling off 20,000 electric vehicles there was no market for electric rentals, high maintenance cost, and collision repairs. The lithium-ion battery powered vehicles were encountering several serious problems. The general public were not in favor of electric vehicles. The electric vehicle rental car business was burdened with a lack of interest in renting electric vehicles. The EV cost was too high. "Electric vehicles have higher price tags, on average, than gas-powered vehicles, although the gap has been narrowing and federal tax credits sometimes exceed the difference." (Domonoske, Copley, 2024) There is practically no used electric car market because if a battery fails the cost of the replacement and installation would be more that the cost of the used car. Manufacturers are losing money on every electric car sold, which is not a good business model. The EPA issued final rules that expected the industry to meet the limits of 56% of new vehicles sales be electric vehicles by 2032. China was controlling the lithium market.

The recharging infrastructure was not adequate. A fully charged vehicle did not have a very long range. The government's drive to transition to electric, zero-emission vehicles was in trouble.

The answer may be as simple as a glass of water. Hydrogen fuel-cell vehicles may become the second evolution of electric vehicles that answer the need for zero-emission vehicles. HFC vehicles have longer range, can be refueled in just minutes, and are better suited for heavy duty vehicles. There were several manufacturers already producing HFC vehicles in 2023, and others were conducting research. The HFC vehicles produce their own electricity to power stacks to drive their wheels.

In this book hydrogen fuel cell vehicles will be compared side by side with the battery powered electric vehicles, how is hydrogen produced, answer the question why hydrogen fuel cells, how they work, how they can be used and describing the advantages and disadvantages of hydrogen fuel cells. An examination of manufacturers research and production plans for HFC vehicles ends with the need for infrastructure to make the hydrogen fuel cells successful.

CONTENTS

1.

COMPARISON of ELECTRIC BATTERY POWER vs. HYDROGEN FUEL CELL POWER

"MANY PEOPLE EITHER DISCOUNT HYDRO-gen fuel-cell propulsion technology or simply don't know much about it. A fuel cell uses hydrogen as a fuel, it combines hydrogen and oxygen to produce electricity and water (and a small amount of heat) as by products." (Smirnov, 2022) "Hydrogen will likely remain a niche technology among cars, pickups, and SUVs, but like diesel has been an alternative to gas. Hydrogen still stands a decent chance of becoming the fuel of choice, vehicles face the same challenges as battery-electric models, including consumer acceptance, fueling infrastructure and cost. Those hurdles are why many expect fuel cells to first enter commercial applications such as trucking with its set routes and destinations" (Wayland, 2024) "Fuel cells for long-distance trucking, heavy-duty work, and power generation in a zero-carbon-emissions future. Fuel cells offer longer range and quicker refueling than battery-electric vehicles, giving them an advantage in industries where every minute of downtime has an impact on the bottom line."

(Tingwall, 2024) "The Biden administration recently proposed new tax guidelines aimed at making it cheaper to produce hydrogen as a less polluting alternatives to fossil fuels." (Hawkins, 2024)

"When it comes to alternative power sources for engines, to the mind of the general public, the fuel cell battery currently lags behind. Yet experts believe that hydrogen fuel cell cars will catch up in the future." (Nils, 2022) "Fewer pollutants, less noise-these are among the many great hopes for electrically powered vehicles. When it comes to electromobility, most people think of vehicles with a large battery that you charge from a wall outlet. Yet there is another propulsion technology that traffic experts are expecting a lot from-including an alternative to long charging times." (Nils, 2022) That alternative is hydrogen fuel cell power. "Fuel cell vehicles at large have the obvious advantage over battery electric vehicles or BEVs in the sense that most offer a better range and can be refueled far quicker. Both are crucial aspects in making cars with newer technology more popular among global buyers." (HT Auto Desk, 2022) Electric cars may be environmentally friendly in terms of emissions, but the mining of lithium and other rare earth materials, plus the energy used in refining, transporting materials, manufacturing of the batteries and shipping may actually be more harmful to the environment. "While the majority of people think that electric cars are environmentally friendly, the truth is that even when they are charging at home, they still release carbon dioxide into the atmosphere. On the other hand, hydrogen fuel cells are completely green." (LaGrada, 2024)

Impact on power grid

Currently there are temporary power grid "brown outs," in various regions of the United States. If millions of electric cars are hooked up to recharge this will increase the burden on our already burdened power grids. Where will the power for recharging electric cars come from? How frequent will the brown outs occur? Hydrogen fuel cell vehicles will not burden our electric grids.

Impact of weather on recharging EVs

Extreme weather affects the performance of batteries. In areas where the temperature exceeds 100+ degrees the batteries of electric cars would be affected. In the upper United States where temperatures often drop below zero lithium-ion vehicle batteries would be affected. The net result, electric vehicle owners might find their cars parked and unusable. Massive snowstorms have hit urban communities resulting in traffic jams where cars are stranded for hours. Gas powered cars have run out of gas. But, what about electric cars running down the battery. Cars are left abandoned only to be towed a couple days later. Owners of gas-powered cars can go get a can of gasoline and fill up on site with enough gas to get started and drive off. In January 2024, in Chicago there were three days in a row where the high temperature for the day was sub-zero. As a result, there were hundreds of EVs stranded because they couldn't be charged and in some cases the charging stations didn't work. Where does the electric vehicle owner go to charge his car?

Infrastructure of charging stations

I noticed that a charging station that had 12 stations for electric cars to charge while shopping at a large, nationally known grocery store stood empty. Apparently electric car owners don't feel that putting their cars on a charger is worthwhile while they are shopping. I noticed at a large mall with over 50 stores there was a charging station for just two electric cars that stood empty. First, two stations for 50 stores are hardly enough. However, charging stations at private businesses has not caught on by 2023.

Recent government action has made billions of dollars available for developing nationwide electric vehicle charging stations. Where will the charging stations be built? Perhaps along Interstate highways, at distances that will be comfortable for the limited distances that an EV can travel on a single charge. Would large urban areas have several charging stations at local gas stations. Will those gas stations be compensated by the federal government for installing the charging stations. How many will be built and how far apart would they be built in urban areas. The question, can enough recharging stations be built to handle future demand? What will be the cost? Will there be long lines of vehicles waiting to hook up? How much will it cost to use government and private charging stations. How much time will be necessary to charge a car? Will shoppers just charge their vehicle for a half hour or an hour just to increase the distance between full charges?

As early as 2004, Former California, Arnold Schwarzenegger vowed to revamp California's highways as "Hydrogen Highways." J.R. DeShazo, director of the UCLA Luskin Center, doubts the hydrogen highways will ever be realized. "If there were stations everywhere, hydrogen would be an obvious solution, he told ABC News. Refueling stations are really expensive and require significant economies of scale to be cost effective and compete with gasoline and electricity." (Korn, 2020) In 2015, Edmunds reported that funding from California's Alternative and Renewable Fuel and Vehicle Technology Program for construction hydrogen refueling stations at an average cost of $1.5 million per station. (Edmunds, 2015) Hyperion, CEO Angelo Kafantaris, plans to build a network of Hyper: Fuel Mobile Stations, to refuel the Hyperion XP-1 hypercar.

Hydrogen fuel cell vehicles will need a massive infrastructure as well. However, hydrogen fuel cells can be fully charged in 9-10 minutes. One expectation is that hydrogen fuel-cell stations will just be additional pumps added to regular gas stations. HFC vehicles can refuel just as fast as gasoline powered vehicles so there would not be long lines waiting to fuel up. Because HFC vehicles will have longer range, there can be a greater distance between refueling stations. As the hydrogen fuel-cell vehicles expand, hydrogen refueling stations will grow together with the rollout of hydrogen fuel-cell vehicles.

There will need to be safety, codes, and standards established for the hydrogen fuel stations. Many of the current hydrogen safety codes and standards are based on regulations of the chemical and aerospace industries. The DOE is coordinating the efforts of codes and standards organizations that ensure the safe use of hydrogen for transportation and stationary applications.

Infrastructure for rural communities

Hydrogen highways are a chain of hydrogen-equipped filling stations located along state and county roads in both urban and rural roads. In time, hydrogen fueling stations will be as common as the current gasoline stations. Hydrogen stations which are not situated near a hydrogen pipeline get supplied by hydrogen tanks, compressed hydrogen tube trailers, liquid hydrogen trailers, or liquid hydrogen tank trucks. Another source of hydrogen fuel would be hydrogen pipelines. Hydrogen pipelines would run from the hydrogen production plant to the point of delivery to vehicles. Electric car owners that live in rural communities will have to travel to get their carts charged and that will take hours of lost time. Hydrogen fuel cell vehicles will have longer driving ranges.

Infrastructure for private residences

Some residences have garages where the electric vehicle can be plugged in overnight. Some homes don't have garages, but the vehicle can be parked close to the house and can be charged from an external power source on the house. In some cases, these home charging stations may require extra cost in installing home charging stations. But what about people that live in multistoried apartment buildings where their cars are parked in an open lot outside. How will those owners be able to charge their cars. If the apartment management-built charging stations rent would be raised to recover the cost. The management could not provide enough charging stations. There would also have to be rules for the use of the charging stations. In many urban communities, cars are parked on the street. How would these cars be charged? People often have to find a on street parking space on street halfway down the block. How would they charge their car? There will be massive numbers of EVs sitting along the curb unable to move.

Hydrogen fuel cell cars will not require extension cords to plug in to recharge the fuel cell. Owners will fuel up at their local gas station so there will be no need for any charging stations built into the garage. For private residences the refueling process will be the same as the owners of gas-powered cars do now.

Vehicle weight

Electric vehicle batteries are not the type that could be bought at big box stores or auto supply dealerships. The lithium-on batteries in an EV are massive. Therefore, they weigh quite a bit and add extra weight to the vehicle. Estimates of the lithium-ion battery is around an additional 1,000 pounds to the car weight. The extra weight causes a loss in operation efficiency. The average electric car has 25 lbs. of lithium, 60 lbs. of nickel, 44 lbs. of manganese, 30 lbs. of cobalt and 200 lbs. of copper. The hydrogen fuel cell vehicle does not have this weight penalty. BMW Sales Chief, Pieter Nota had this to say regarding larger SUVs. "We see that hydrogen fuel cell technology is particularly relevant for larger SUVs." (Hydrogen Central, 2022)

Battery replacement

Most electric cars produced around 2023 had an estimated life of ten years. Whether the original owner keeps the car for ten years or passes it on to other owners the life cycle will eventually be met, and a new battery will be necessary. For the owner of a used car may find the cost to replace the battery is greater than what they paid for the used car. This will impact the ability to sell used EVs. If the original owner keeps the car for more than ten years, then the original owner will face the huge replacement battery cost. This must be factored in before the original purchase is made.

Lithium-ion batteries can't be recycled.

Ironically, the transformation to EVs was designed to save the environment from fossil fuels. The solution to one problem can lead to another problem. The electric car battery contains lithium, nickel, manganese, cobalt, and copper, plus aluminum, steel, and plastics. A major environmental problem is caused by lithium-ion batteries because they cannot be recycled. In contrast the hydrogen fuel cell only has a platinum membrane which may be replaced in newer versions with a ceramic membrane, eliminating the use of rare metals. Eventually the EV batteries will be dumped into a massive landfill and then buried under tons of earth. Three or four hundred years later dangerous chemicals from the lithium-ion batteries will seep into the ground and eventually lead to pollution of the earth from dangerous chemicals. Hydrogen fuel cells do not produce this problem.

Lithium-ion powered heavy duty vehicles are not suited for heavy payloads

The batteries needed for trucks with heavy payloads would be huge. Therefore, manufacturers have determined that the hydrogen fuel cell is better suited for heavy payloads. "Whether electrifying an 18-wheel truck, a coach, a city bus or some similar sort of long-endurance fleet vehicle, hundreds of kilowatt-hours of energy need to be stored onboard. The problem here is that commercially available lithium-ion cells typically contain only 0.015 kWh/kg; even the very best, from r&d labs, rarely hold more than 400Wh/kg. Furthermore, given the amount of weight to be carried and the levels of vibrational and environmental harshness likely to be endured, the cells must be housed in packs large enough to house extensive protection systems to prevent any damage to them. That piles on the weight though forcing EV engineers to add more and more packs to ensure their vehicles have enough onboard energy to satisfy their range conditions." (E-Mobility, 2023) In simple terms the batteries for heavy duty vehicles would have to be massive. John Warner, chief customer officer for American Battery Solutions had this to say. "As the vehicle weight increases, energy demand increases. You also have to build in some contingency for hot or cold weather, and some reserve." (Park, 2022)

Return on investment

Currently electric cars cost much more than fossil fuel driven vehicles. The question is how long does it take to make up the difference in cost? "Some EVs can cost more to purchase than a similar gas-powered car, and certain maintenance expenses (such as battery replacement, if needed) are also higher. Owners may also need to install a charging station at home, which adds to the upfront cost." (Discover, 2023)

At the Chicago Auto Show in 2023 one of the manufacturers had a gas model on display and the sticker on the windshield said $35,000. Sitting right alongside was an all-electric, exactly the same model with the same accessories. The sticker on the windshield read $58,800, a difference of $23,800. How long would it take to get a return on the investment?

Times change but buyers don't. Recent surveys in a study by Jerry has shown that one third of American buyers say they will never buy and electric vehicle. The same survey also showed that 60% of American buyers didn't foresee buying an electric car in the next 6 years. In the short-term, drivers will likely not find the immediate (Return on Investment) in buying an EV. (Tengler, 2022) "This does not bode well for governments like the U. S. government whose infrastructure plans announced in December assume an ambitious target of 50% of sales by 2030 being electric to support the business case." (Tengler, 2022)

Environmental Impact

Are electric batteries actually bad for the environment? "But between manufacturing, charging and recycling them, these batteries may not be as green as they seem." (Sheldon, 2022) A major source of lithium are the salt flats of Chili, Argentina, and Bolivia. Immense amounts of water is pumped down into the salt flats to bring the lithium rich salt water to the surface, where it is left to evaporate. This process runs the risk of water contamination in the area. Some lithium is mined in the form of ore which requires processing to make it usable. One of the elements needed is cobalt. The process of mining for cobalt can produce hazardous byproducts that can toxify the environment. Cobalt mines often produce sulfuric acid which when exposed to air and water can create damage to the environment that can last for hundreds of years according to the United Nations. " Research by the European Environmental Agency found batteries alone account for 10% to 75% of the energy and 10% to 70% of the greenhouse gas emissions resulting from the entire production of the vehicle." (Sheldon, 2022) The same EEA study found production of electric cars emits between 1.3 and 2 times the amount of greenhouse gases that that of internal-combustion vehicles." (Sheldon, 2022) In most cases the electricity they get for the charging stations likely is produced by fossil fuels. 99% of lead car batteries are recycled but most lithium batteries end up in landfills, where their hazardous components can leak into the soil and groundwater. Considering the above information, hydrogen fuel cell technology appears to be a better choice for the environment.

2.
HOW is HYDROGEN PRODUCED?

What is hydrogen?

"Few chemicals carry as much hope and aspiration as hydrogen. Over the last few years the first element in the periodic table has gone from a global buzzword too one of the most promising routes to decarbonizing industry power generation and transport." (Willige, 2022) Hydrogen is the simplest element on earth-it consists of only one proton and one electron—and it is an energy carrier, not an energy source. Hydrogen can store and deliver usable energy. Hydrogen can be produced from fossil resources, biomass/waste and by splitting water. "Hydrogen is the lightest and most common element in the cosmos. Its atomic number is one. In its elemental state, hydrogen is rare. But it is one of the components of water and vital to life." (Uses of hydrogen, 2020) Hydrogen is a non-polluting alternative fuel that is considered as an alternative to for transportation fuel. "Hydrogen fuel cells directly convert the chemical energy in hydrogen to electricity and release water and useful heat as by-products. Hydrogen fuel cells are pollution-free and have greater efficiency than traditional combustion technology." (Walker, 2020) Hydrogen can be produced through low-carbon pathways using diverse, domestic resources—including fossil fuels, such as natural gas and coal, coupled with carbon capture and storage: through splitting of water using nuclear energy and renewable energy sources, such as wind, solar, geothermal, and hydro-electric power; and from biomass through biological processes. "The raw materials to produce hydrogen are abundantly present on earth, minimal land is required for production, the substance has an extremely high gravimetric energy content, and the only by-products are heat and water. This makes hydrogen a highly sustainable fuel and energy carrier when produced using green energy. This sustainable form of hydrogen is known as green hydrogen." (Demarco, 2024) "Hydrogen fuel can be produced through several methods. The most commonly used methods today are natural gas reforming (a thermal process), and electrolysis of water, which sees electricity running through water to separate the hydrogen and oxygen atoms. The electricity used can be generated by wind, solar and hydro sources." (Youd, 2021).

Hydrogen has three isotopes, protium, deuterium and tritium. Protium has one proton and no neutrons. Deuerium has one proton and one neutron and is more massive and heavier than protium and is therefore called heavy hydrogen. The third isotope, tritium, has one proton and two neutrons and is also regarded as heavy hydrogen. Deuteriumis a stable isotope which means it is not radioactive. There is a natural abundance of deuterium in the ocean. The name deuterium comes from the Greek for deuteros which means second, refers to the two particles, a proton and neutron. Sometimes deuterium is referred to as deuteron or deuton. Both tritium and deuterium form stronger bonds than protium.

Tritium is also known as H-3. Tritium has a half-life that ranges from 7-14 days. A half-life is defined as the time it takes for half a given amount of material to decay. Just as deuterium, the name tritium comes from the Greek tritos which means third. Tritium decays into Helium-3 via beta particle emission. When a substance decays into another substance the process is called natural transmutation. Unlike deuterium, tritium is radioactive. Tritium is mainly found in liquid form but can exist as an odorless and colorless gas. Trace amounts of tritium occur naturally on Earth. External exposure to tritium is not very dangerous because tritium emits a low energy beta particle. This beta particle radiation is so weak that it cannot penetrate the skin. However, if ingested, tritium can be dangerous. Tritium has been used as a radioactive tracer in biological and environmental stuies. Tritium has served as a label in chemistry experiments. Tritium has also been used as a radio luminescent light source for watches and instruments. Tritiated water, when tritium combines with oxygen, can be used as a tool to monitor the hydrologic cycle and to map ocean currents.

Hydrogen Fuel Basics

"Hydrogen is a clean fuel that, when consumed in a fuel cell, produces only water. Hydrogen can be produced from a variety of domestic resources, such as natural gas, nuclear power, biomass, and renewable power like solar and wind. These qualities make it an attractive fuel option for transportation and electricity generation applications. It can be used in cars, in houses, for portable power, and in many more applications." (Office of Energy Efficiency, 2020)

Hydrogen is an energy carrier that can be used to store, move, and deliver energy produced from other sources. Pressurized hydrogen interacts with Oxygen in the air to create electricity through a chemical reaction. Today, hydrogen fuel can be produced through several methods, The most common methods today are natural gas reforming, a thermal process, and electrolysis. Other methods include solar-driven and biological processes.

Today, hydrogen fuel can be produced through several methods. The most common methods today are natural gas reforming (a thermal process), and electrolysis. Other methods include solar-driven and biological processes.

"As the population increases the carbon dioxide emission increases at the same time. However, the rise in carbon dioxide has harmful consequences not only on our planet but also for the future for the next generations. Therefore, it needs to find alternative to the use of fossil fuels for the energy production." (Krishnakumar, et al. 2020) "Energy security and climate change are very popular issues these days and their imperatives require a large-scale substitution of petroleum-based fuels as well as improved efficiency in energy conversion." (Rabenstein, Hacker, 2008) With environmental regulations becoming more stringent, hydrogen becomes an alternative fuel possibility.

Thermal Processes

Thermal processes for hydrogen production typically involve steam reforming, a high-temperature process in which steam reacts with a hydrocarbon fuel to produce hydrogen. Many hydrocarbon fuels can be reformed to produce hydrogen, including natural gas, diesel, renewable liquid fuels, gasified coal, or gasified biomass. Today about 95% of all hydrogen is produced from steam reforming of natural gas.

The steam reforming process, after desulfurization, utilizes the natural gas feed by combing it with steam. The mixture is then passed through a reactor filled with catalysts, resulting in a mixture of carbon monoxide, carbon dioxide, and hydrogen. The steam reforming process is industrially mature. As a result, there is no unknown in the process as it is very well characterized. This reduces down time and risk and ensure the safety of the workers." (Dotzenrod, et al, 2020)

Electrolytic Processes

Water can be separated into oxygen and hydrogen through a process called electrolysis. Electrolytic processes take place in an electrolyzer, which functions much like a fuel cell in reverse-instead of using the energy of a hydrogen molecule, like a fuel cell does, an electrolyzer creates hydrogen from water molecules. Electrolysis produces no greenhouse gas emissions. Electrolysis does not require complex equipment. Similar to electrolysis, photoelectrochemical splitting involves splitting water into hydrogen and oxygen by using sunlight and specialist semiconductors for the process. "The hydrogen itself can be produced by running this process in reverse, which is called electrolysis. Running an electrical current through water separates the H2O into hydrogen and oxygen. More commonly, however, hydrogen is produced at scale from natural gas in a process called steam-methane reforming, in which high-temperature and high-pressure steam is combined with natural gas to create hydrogen." (Thompson, 2020)

Solar-Driven Processes

Solar-driven processes use light as the agent for hydrogen production. There are a few solar-driven processes, including photobiological, photoelectrochemical, and solar thermochemical Photobiological processes use the natural photosynthetic activity of bacteria and green algae to produce hydrogen. Photoelectrochemical processes use specialized semiconductors to separate water into hydrogen and oxygen. Solar thermochemical hydrogen production uses concentrated solar power to drive water splitting reactions often along with other species such as metal oxides.

Biological Processes

Biomass gasification is a mature technology pathway that uses a controlled process involving heat, steam, and oxygen to convert biomass to hydrogen and other products, without combustion. Biological processes use microbes such as bacteria and microalgae and can produce hydrogen through biological reactions. Green algae or other microbes consume sunlight and water and produce hydrogen in the process. In microbial biomass conversion, the microbes break down organic matter like biomass or wastewater to produce hydrogen, while in photobiological processes the microbes use sunlight as the energy source.

"Biomass gasification or renewable liquid reforming: a form of biomass (be it a solid, for example plant-based biomass, or a liquid for example ethanol) can be reached with very high temperature steam to produce hydrogen and carbon dioxide. This is a similar process to SMR, but the carbon dioxide here is from a renewable energy source that originated within the Earth's atmosphere (e.g. entering plants via photosynthesis) and thus the overall carbon footprint is much lower. The equipment required would mean that more centralized production (e.g. in a power plant) would be required."(Perry, 2018)

"Biomass is a renewable energy source with an enormous potential to generate heat, power, and biofuels. It is, therefore, one of the extensively explored research areas in sustainable energy field. Biomass gasification is commonly employed thermal chemical route to get high-value end products and potential applications. Not only low-cost feedstocks can be employed in the biomass gasification process, but many organic wastes such as municipal solid wastes can also be treated for usable energy generation. Cutting-edge, innovative, and cross-effective gasification techniques with high efficiencies are the essential criteria for the expansion of this technology." (Sikarwar, Zhao, 2017) The significance of this type of biomass gasification is that almost anything can be a fuel source for gasification. Examples of the variety of what can be included as biomass include forest biomass, grasses, cultivated crops, algae, water plants, biosolids, sewage, landfill gas, manure, leaves, demolition wood, sawdust and waste oil. That would go a long way towards improving the environment. As biomass gasification is a mature technology, feedstock costs and lessons learned from commercial demonstrations will determine its potential as a viable pathway for cost-competitive hydrogen production.

Pyrolysis is the gasification of biomass in the absence of oxygen. In general, biomass does not gasify as easily as coal, and its produces other hydrocarbon compounds in the gas mixture existing the gasifer; this is especially true when no oxygen is used. As a result, typically an extra step must be taken to reform these hydrocarbons with a catalyst to yile a clean syngas mixture of hydrogen, carbon monoxide, and carbon dioxide. Then just as in the gasification process for hydrogen production, a shift reaction step (with steam) converts the carbon monoxide to carbon dioxide. The hydrogen produced is then separated and purified.

"Syngas is the main end product of biomass gasification and is an important source of environmentally friendly fuels and chemicals. It is a suitable feedstock for the electricity generation. Gasoline, diesel, and other chemicals can be synthesized employing Fischer-Tropsch (FT) process. Hydrogen-rich syngas can be obtained from gasification via the water-gas shift reaction, which has numerous applications including in fuel cells." (Sikarwar, Zhao, 2017)

Thermochemical production or high temperature water splitting when extreme high temperature steam produced by nuclear power would be used to split water at around 1800 degrees Celsius yielding hydrogen in the process.

Steam Reforming

Hydrogen can be produced commercially on a large scale mainly by steam reforming. "Methanol, ethanol, ammonia, gasoline, and natural gas are some possible sources of hydrogen for fuel cells. In addition, petroleum distillates, liquid propane, oil, gasified coal and even gas from landfills and wastewater treatment plants can also be processed to supply hydrogen." (Vaidya, Rodrigues, 2006) Of these, ethanol seems to be the most reasonable. Ethanol is a promising candidate for a renewable source for hydrogen production. Ethanol is currently used in fossil fuels and is important economically to corn farmers across the country. Ethanol can be produced from agricultural crops, biomass fermentation, woodwork and food residues. Ethanol can be converted to methane, which is converted to hydrogen under increased temperatures. Steam reforming achieves the highest hydrogen-yield. "Development of hydrogen energy technologies using novel and renewable power sources represents a potential energy solution for the future. One of priority choices focuses on the development of mobile power generation units based on fuel cells. Promising potential of this approach relies on high efficiency of chemical-to-electrical energy transformation, environmental safety of power generation process, possibility to design efficient reliable low-power systems." (Sntnikov, et al, 2012)

Natural Gas Reforming

Natural gas reforming is an advanced and mature production process that builds upon the existing natural gas pipeline delivery infrastructure. Today, 95% of the hydrogen produced in the United States is made by natural gas reforming in large central plants. This is an important technological pathway for near-term hydrogen production. Natural gas contains methane that can be used to produce hydrogen with thermal processes, such as steam-methane reformation and partial oxidation. Most hydrogen produced today in the United States is made via steam-methane reforming. Steam reforming is endothermic—that is, heat must be supplied to the process for the reaction to proceed. In the water-shift reaction the carbon monoxide and steam are reacted using a catalyst to produce carbon dioxide and hydrogen. Steam reforming can also be used to produce hydrogen from other fuels, such as ethanol, propane, or even gasoline. Another form of reformiong is a partial oxidation. In partial oxidation, the methane and other hydrocarbons in natural gas react with a limited amount of oxygen that is not enough to co0mpletely oxidize the hydrocarbons to carbon dioxide and water. Partial oxidation is an exothermic process—gives off heat. This process is typically much faster than steam reforming and requires a smaller reactor vessel.

Reforming low-cost natural gas can provide hydrogen today for fuel cell electric vehicles (FCEVs) as well as other applications. Over the long term, DOE expects that hydrocarbon production from natural gas will be augmented with production from renewable, nuclear, coal, and other low-carbon domestic energy resources. Petroleum use and emissions are lower than for gasoline-powered internal combustion engine vehicles. The only product from an FCEV tailpipe is water vapor but even with the upstream process of producing hydrogen from natural gas as well as delivering and storing it for use in FCEVs, the total greenhouse gas emissions are cut in half and petroleum is reduced over 90% compared to today's gasoline vehicles.

Gasification Processes

The United States Department of Energy's Office of Fossil Energy, through the Gasification Systems Program ,is developing flexible, innovative, resilient, and transformative modular designs for converting diverse types of US domestic coal and coal blends with biomass, municipal solid waste (MSW), and waste plastics into clean synthesis gas to enable the low-cost production of electricity, high-value chemicals, hydrogen, transportation fuels, and other useful products to suit market needs, combined with negative emission of greenhouse gases technologies.

The Unites States has an abundant, domestic resource in coal. The use of coal to produce hydrogen for the transportation sector can reduce America's total energy use and its reliance on imported petroleum while helping create jobs through the creation of a domestic industry. The production of hydrogen from coal also offers environmental benefits when integrated with advanced technologies in coal gasification, power production, and carbon capture, utilization, and storage. The integration of these technologies facilitates the capture of multiple pollutants such as sulfur oxides, nitrogen oxides, mercury, and particulates, as well as greenhouse gases such as carbon dioxide. When hydrogen is used in efficient fuel cell vehicles, emissions from the transportation sector can be nearly eliminated. In coal gasification, coal, is mixed with steam and oxygen to produce syngas, a mixture of carbon monoxide, hydrogen, and methane. The hydrogen obtained from coal gasification can be used for a variety of purposes. Hydrogen is produced by first reacting coal with oxygen and steam under high pressure and temperatures to form synthesis gas.

"Biomass gasification means incomplete combustion of biomass resulting in production of combustible gases consisting of Carbon monoxide (CO), Hydrogen (H2), and traces of Methane (CH4). This mixture is called producer gas. The production of generator gas (produce gas) called gasification, is partial combustion of solid fuel (biomass) and takes place at temperatures above 10000C. The reactor is called a gasifier. Since there is an interaction of air or oxygen and biomass in the gasifer, they are classified according to the way air or oxygen in introduced in it. There are three types of gasifiers Downdraft, Updraft, and Crossdraft."(Thorat, 2020) "In the updraft gasifier, biomass feedstock is fed from the top end, and gasifying agent comes from the bottom, and in downdraft both fuel and gasification agent are fed from the top with the fuel coming in from a lock-hopper."(Sikarwar, Zhao, 2017) The nature of the fuel involved in biomass gasification suggests that it would not produce sufficient H2 to serve as fuel for fuel cell vehicles.

How is hydrogen used?

Because hydrogen is part of so many bio-molecules like carbohydrates, proteins, and fats. Hydrogen has nearly three times the energy content of gasoline. Therefore, hydrogen has many different uses. Nearly all of the hydrogen consumed in the United States is used by industry for refining petroleum, treating metals, producing fertilizer, and processing foods. U.S. Petroleum refineries use hydrogen to lower the sulfur contents of fuels. Hydrogen combines readily with sulfur.

"Uses of hydrogen can be listed in terms of its plain gaseous uses and also elemental applications. But in both ways, it funds district uses which are not replaceable by other elements or gases. Hydrogen is available as natural gas in nature. But in pure form, it is highly inflammable. Further, it is also available in the form of isotopes like deuterium, tritium, ets. All these properties of hydrogen contribute towards its uses."(Ranga, 2015)

Gas welding

Welding industries use hydrogen for welding torches. "A common use of hydrogen gas is in the gas welding process. It is used in this type of welding to generate a high temperature of 4000 degrees C. this high temperature leads to the melting of metals and thereby joining the broken surfaces. Besides generating heat, hydrogen also acts as a shielding gas. Since metals at high temperature are very reactive, hydrogen prevents them from reacting with other elements like nitrogen, carbon during the process. This process is also called as atomic hydrogen welding." (Ranga, 2015)

Automobile fuel

BMW set a land speed record with a hydrogen fueled internal combustion engine. The hydrogen burns with oxygen to provide power while producing an exhaust of water vapor. The combustion of hydrogen fuel yields plenty of energy. This technology is still being developed to replace petroleum burning engines.

In 2020 Hyperion announced plans for a hydrogen powered supercar.

In petroleum refinery

"Hydrogen gas widely used in petroleum industry to remove sulfur content. Besides, it is also used for hydroisomerization wherein normal paraffin is converted to isoparaffin. Dearomatisation to convert aromatic to cycloalkanes. Hydrocracking to break long-chain hydrocarbons to shorter ones."(Ranga, 2015)

Ultraviolet lamps

"Hydrogen gas in the form of dueterium is used in deuterium arc lamps. These lamps are used to produce ultraviolet light. These lamps have a tungsten filament which generates heat. The heat in the bulb excites the deuterium atoms (H2) which produce light in the ultraviolet region. These lamps are used explicitly in the spectroscopic analysis in labs."(Ranga, 2015)

As a reducing agent

Hydrogen is required as a reducing agent in chemical industries. "Hydrogen acts as both an electro-negative and electro-positive element. This electro positive nature makes it useful in redox reactions. Addition of hydrogen is termed as reductive while removal is called oxidation."(Ranga, 2016)

In chemical analysis

"Hydrogen is used in various methods of chemical analysis. It is used in Hydrogen electrode for titrations by potentiometry. Here hydrogen gas is passed in a constant speed at one atmosphere pressure into the reference electrode. The purity of the hydrogen can affect the analysis on these potentiometric titrations. These methods include especially the atomic absorption spectroscopy."(Ranga, 2015)

In structural identification

"Many compounds have complex chemistry. Different methods are used to study the structure and the nature of bonds between atoms and the molecules. Nuclear magnetic resonance (NMR) is one of the techniques which exploits the proton character to define the molecular structure. The technique is also called as proton NMR. The presence of hydrogen in many biomolecules is utilized for the purpose. Hence the presence of hydrogen helps in structural identification." (Ranga, 2016)

Gas chromatography

Hydrogen is one of the gases which can be used as a carrier gas in gas chromatography. "Hydrogen, as a carrier gas for GC, can be generated at low pressure on a local basis to provide significant safety and convenience compared to the use of tank gas." (Froelich, 2007) Historically, helium and nitrogen have been used as carriers. A major benefit of hydrogen over helium and nitrogen is the reduction time for the process and allows for the operation at lower temperatures. "The use of hydrogen as a carrier gas for GC provides more rapid separations than nitrogen, with a minimum loss in chromatographic efficiency. The mode of supplying hydrogen is via the electrolysis of water. A hydrogen generator creates a steady stream of gas at low pressure and stores a very small quantity of the actual gas, so that safety issues due to the potential of an explosion are dramatically minimized. In addition, the hydrogen generator is more convenient than tank gas, requires essentially no maintenance, and reduces the cost of hydrogen relative to the use of tank gas. A single hydrogen generator can provide the carrier gas for several GC systems as well as the gas needed for detectors."(Froehlich, 2007)

Deuterium for electricity generation

Hydrogen has two isotopes, deuterium and tritium. Deuterium is used to make heavy water which is used as a coolant in a nuclear reactor. Deuterium is also used as a moderator to slow down the neutrons in a reactor. Deuterium is also used as a tracer in biological research. Deuterium serves as great fuel for experimental fusion reactors, amateur fusors and neutron generators. Deuterium is widely used in prototype fusion reactors and the reactors use in military, industrial and scientific fields.

In balloons

One of the earliest uses for hydrogen was in flying balloons. However, because of hydrogen's explosiveness, hydrogen was replaced with Helium which was much more stable. Because hydrogen is light, scientists are able to use it with weather balloons. Weather balloons are fitted with equipment to record information necessary to study the climate, predicting hurricanes and local daily, weekly, and monthly weather systems. Initially hydrogen was used a fuel for hot air balloon flight. However, because of hydrogen's explosiveness it has been replaced with more stable Helium.

Hydrogen balloons were used as airships in World War I.

Rocket fuel for space programs

Hydrogen in liquid form has been used as rocket fuel. The National Aeronautics and Space Administration (NASA) began using liquid hydrogen in the 1950s as a rocket fuel, and NASA was one of the first to use hydrogen fuel cells to power the electrical systems on spacecraft. Hydrogen has a higher specific impulse. A higher specific impulse means that liquid hydrogen is able to produce more thrust for a longer time and has better fuel efficiency. Liquid hydrogen's high specific impulse means it can make rockets go farther and faster while using less fuel. Hydrogen is a powerful propellent. "Hydrogen, the fuel for the main engines, is the lightest element and normally exists as a gas. Gases—especially lightweight hydrogen—are low—density, which means a little of it takes up a lot of space. To have enough to power a large combustion reaction would require an incredibly large tank to hold it—the opposite of what's needed for an aerodynamically designed launch vehicle." (Harbaugh, 2016) This why liquid hydrogen is used.

Nuclear reactors

Heavy hydrogen in the form of an isotope known as Deuterium is used for nuclear fusion reaction in nuclear reactions. Hydrogen is used as a moderator in a nuclear reaction. Hydrogen works well as a moderator because its mass is almost identical to a neutron. This means that one collision will significantly reduce the speed of the neutron because of the law of conservation of energy and momentum. Nuclear power plants can produce hydrogen in a variety of methods that would greatly reduce air emissions while taking advantage of the constant thermal energy and electricity it reliably produces.

Mass destructive bombs

The other hydrogen isotope, Tritium, has been used in the development of a powerful weapon of mass destruction. Tritium, is generated in nuclear reactions. Tritium a radioactive isotope, is used in making the H-bomb. An atomic bomb uses either uranium or plutonium and relies on fission, a nuclear reaction in which a nucleus or an atom breaks apart into two pieces. To make a hydrogen bomb, one would still use uranium or plutonium as well as two other isotopes of hydrogen, called deuterium and tritium. Isotopes are different forms of an atom with a different number of neutrons.

Hydrogen peroxide

When a child would fall and scrape his knee the parent would put hydrogen peroxide on the scrape. Bubbles would form on the skin. The peroxide was reacting with enzymes in the skin to clean the wound. The bubbles were oxygen being released by the reaction. "Hydrogen peroxide is used in many ways. First and foremost it is used for medication. It is included in most first aid kits. It is primarily used for treating wounds, and cuts. Peroxide is also a toenail fungus disinfectant. Hydrogen peroxide can be diluted in water. It can kill bacteria and germs if used as whitewash. The same element can be used for teeth whitening and canker sores treatment."(Uses of, 2020) Hydrogen peroxide has also been used to clean stains in clothing, a bleaching agent for cleaning and for pest control in your garden.

Multiple other uses

Hydrogen is used in methanol production, bioscience as an isotopic label, and the fertilizer and paint industries. Methalcohol is used in inks, varnishes, and paints. In addition, hydrogen is used in the chemical and food industries. Chemical industries use hydrogen to produce hydrochloric acid. Hydrogen is used in hydrogenated vegetable oils such as margarine and butter, by combining vegetable oils with hydrogen. Using nickel as a catalyst, solid fat substances are produced in the petrochemical industry. Hydrogen is also required for crude oil refinements. Hydrogenating agents are used to convert unsaturated fats to saturated fats. Electrical generators use hydrogen as a rotor coolant. Hydrogen has also been used to purify mined tungsten.

In 2019 Mercedes-Benz and Rolls-Royce announced an agreement for Rolls-Royce and Lab1886, a lab within Mercedes -Benz to begin a pilot project to develop vehicle fuel cells for stationary power generation. The agreement was to begin in 2020. "Over the coming months, Rolls-Royce will develop an integrated MTU solution for sustainable off-grid generation of continuous and emergency power using vehicle fuel-cells, focused on safety-critical applications such as data centres. The pilot project will also include the construction of an emergency power plant for Rolls-Royce's data centre in Friedrichschafen. The plant will be based of fuel cell modules built by Mercedes-Benz Fuel Cell GmbH." (Burgess, 2019)

Hydrogen Fuel Cells

Fuel cells can be used in a wide range of applications, providing [power for industrial/1commercial/residential buildings, and long term energy storage for the grid in reversible systems. Hydrogen fuel cells have been used in spacecrafts, for remote weather stations and even in submarines.

What is a Hydrogen Fuel Cell?

"In 1839, the first fuel cell was conceived by Sir William Robert Grove, a Welsh judge, investor, and physicist. He mixed hydrogen and oxygen in the presence of an electrolyte and produced electricity and water. The invention, which later became known as a fuel cell, didn't produce enough electricity to be useful." (Bellis, 2019) "In 1889, the term "fuel cell," was first coined by Ludwig Mond and Charles Langer, who attempted to build a working fuel cell using air and industrial coal gas, Another source states that it was William White Jaques who first coined the term "fuel cell." Jaques was also the first researcher to use phosphoric acid in the electrolyte bath." (Bellis, 2019) So why fuel cells? What's so great about the use of fuel cells. "Maybe the answer to 'What's so great about fuel cells?' should be the question 'What's so great about pollution, changing the climate or running out of oil, natural gas, and coal?' As we head into the next millennium, it is time to put renewable energy and planet-friendly technology at the top of our priorities." (Bellis, 2019)

A fuel cell uses the chemical energy of hydrogen or other fuels to cleanly and efficiently produce electricity. "Hydrogen fuel cells combine hydrogen and oxygen, creating nothing but pure water as exhaust. By converting the chemical energy stored in the gasses into electrical energy, the energy can be harnessed to power electric drive motors, temporary storage batteries, or a variety of other end applications." (wha, 2023) A fuel cell is any device that produces electricity from a variety of fuels or energy carriers. Hydrogen fuel cells produce electricity by combining hydrogen and oxygen atoms. The hydrogen reacts with oxygen across an electrochemical cell similar to that of a battery to produce electricity, water, and small amounts of heat. "Using hydrogen fuel cells to create electricity is a relatively new process that allows us to fuel electrical devices in a much cleaner way. Depending on the method used to produce the initial hydrogen gas, vehicles of the future could then produce near zero emissions as the only byproducts of generating energy in this way are water and heat. However, to power something as large as a car, the hydrogen cells need to be bunched together as a "stack." This fuel cell stack shows a large amount of electricity to be produced as long as there is a supply of hydrogen available to use." (Viessmann, 2020) The question remains how efficient hydrogen fuel cells and how much hydrogen are is needed in order to get a usable electricity output.

There are several types of fuel cells including: proton-exchange membrane fuel cells (PEMFGs), phosphoric acid fuel cell (PAFC), Solid acid fuel cell (SAFC), Alkaline fuel cell (AFC), High-temperature fuel cells, solid oxide fuel cell and molten-carbonate fuel cell (MCFC). Of the many types of fuel cells, they all consist of an anode, a cathode, and an electrolyte that allows ions, often positively charged hydrogen ions (protons), to move between the two sides of the fuel cell. Fuel cells are classified by the type of electrolyte they use and by the differences in startup time. In the proton-exchange membrane design a proton-conducting polymer membrane contains the electrolyte solution that separates the anode and cathode. The phosphoric acid fuel cell uses a non-conductive electrolyte to pass positive hydrogen ions from the anode to the cathode. The solid acid fuel cell uses a solid acid material as the electrolyte. The alkaline fuel cell was used as the primary source of electrical energy in the Appollo space program. Solid oxide fuel cells use a solid material, most commonly a ceramic material called yttria-stabilized zirconia, as the electrolyte. The molten carbonate fuel cells require a high operating temperature. The molten carbonate fuel cell is capable of converting fossil fuel to hydrogen-rich gas in the anode, eliminating the need to produce hydrogen externally.

3.
WHY HYDROGEN FUEL CELLS?

SUBITH VASU A PROFESSOR IN UNIVERSITY Central Florida's Department of Mechanical and Aerospace Engineering offer this answer to the question; Why hydrogen? "Hydrogen is a simple fuel that can solve both the energy and water crisis in the world, while at the same time help reducing the greenhouse gas impact." (Lawson, 2021) Although the focus of this book is the application of hydrogen fuel cell power systems in vehicles is not the only application for hydrogen fuel cells. Hydrogen fuel cells may someday provide the energy to heat our homes, power our electric grid and many other electric motorized applications. "For energy providers wanting to maximize their return on investment from renewables, provide continuous power, balance the grid over long durations, and provide zero-emission power to their customers, hydrogen fuel cells are the answer. Hydrogen fuel cells are a competitive solution for distributed power generation and critical infrastructure backup power." (Ballard, 2021) 'Described by the International Energy Agency as a 'versatile energy carrier,' hydrogen has a diverse range of applications and can be deployed in sectors such as industry and transport." (Frangoul, 2021) Fuel cells can operate at higher efficiencies than combustion engines, lower or zero emission, no carbon emissions, it emits only water as an exhaust, and they are quiet during operation. "Hydrogen has the potential to be the next fuel source to cross over and be widely adopted in the consumer market, especially as it has been proven that 'it can be done' from a technological perspective." (McEvoy, 2021) Subaru CEO Tomomi Nakamura raised this question regarding the need for an alternative energy system for vehicles. "As no one knows what the future will really look like? Nakamura said, 'The question right now is whether it's really the right move to eliminate the internal combustion engine entirely." (Davis, 2022)

A fuel cell uses the chemical energy of hydrogen or other fuels to cleanly and efficiently produce electricity. If hydrogen is the fuel, the only products are electricity, water, and heat. Fuel cells are unique in terms of the variety of their potential applications: they can use a wide range of fuels and feedstocks and can provide power for systems as large as a utility power station and as small as a laptop computer.

"Hydrogen fuel cell systems are a reliable zero emission power solution for backup, standby, and peak shaving stationary applications. The fuel cell system converts hydrogen into cost-effective electricity that may be used onsite by the customer or sold back to the grid. Key attributes of Ballard's megawatt-scale fuel cell power generation systems."(Ballard, 2021) Among the advantages are smaller space in comparison to other clean energy technologies, a highly efficient and flexible system for cold start and rapid ramp up, improved grid stability greater energy independence and long duration energy storage at lower cost than battery banks.

The interest in hydrogen as a transportation fuel is based on its potential for domestic production and use in fuel cells for high efficiency zero-emission electric vehicles. A fuel cell is two to three times more efficient than an internal combustion engine running on gasoline. Hydrogen use in vehicles is a major focus of fuel cell research and development.

Environmental concerns over the use of fossil fuels have caused automotive manufacturers to look for alternative sources of power. Electric vehicles have gotten the early jump on the market, but hydrogen fuel cell power has shown promise. That promise is the focus of this book. Hydrogen fuel cell vehicles are environmentally friendly because they produce no direct pollution emissions. Hydrogen fuel cell vehicles address two major environmental concerns, fossil fuels and carbon dioxide emissions. "A decade ago, it seemed like there were two potential contenders to replace fossil fuel for personal transportation—electricity and hydrogen. The hydrogen option had a lot going for it. You could fill your car up just like fossil fuel, but instead of noxious gases coming out of the tailpipe, the exhaust would just be pure water vapor. It sounded like the perfect step forward towards a greener future where we could carry on using our vehicles as before, only without the environmental downsides. Compared to waiting around for an EV's battery to recharge, hydrogen appeared to be much more convenient option." (Morris, 2020)

"The potential for fuel cell technology, and of hydrogen for energy storage, is beyond question. Hydrogen plays a major role in the discussion of ways in which to reach worldwide climate targets. It is one of the ways in which the range of fuels in the transport sector can be extended in a climate-friendly manner: because it is particularly with the help of hydrogen generated from renewable energy that harmful CO_2 emissions can be significantly reduced. Operation of a hydrogen-powered fuel cell vehicle causes neither pollutant nor CO_2 emissions." (Mercedes-Benz, 2020)

The history of hydrogen fuel cell cars

"The first hydrogen car invented was not a fuel cell vehicle (FCV) but rather an internal combustion engine. Swiss inventor Francois Isaac de Rivaz in 1807 designed the first 4-wheel protype that ran on hydrogen and oxygen gas. The hydrogen gas was contained in a balloon and the ignition was an electrical Volta starter." (AutomoStory. 2022) 1847 William Grove invented a working fuel cell that converted chemical energy of hydrogen and oxygen into electricity. In 1860, Frenchman Etienne Lenoir built a three-wheel 1 cylinder, 2 stroke engine in his Hippomobile. In 1933, Norway's Norsk Hydro power company converted a truck using a ammonia reformer to produce hydrogen in an internal combustion engine. "Leap forward to 1941 and World War II. Out of dire need Russia's Boris Shelishch converted 200 GAZ-AA trucks to run on hydrogen. Germany's Nazi army had surrounded Leningrad, Russia so Shelishch converted the trucks to run on hydrogen gas. Which burned cleaner and ran longer than those which had run on gasoline." (AutomoStory, 2022) In 1959, Harry Karl Ihrig, working for Allis-Chalmers, created a hydrogen fuel cell powered farm tractor. In 1966 Roger Billings converted a Ford Model A to run on hydrogen and General Motors created the Electrovan which was a fuel cell vehicle. "The late 2000s and early 2010s were a boom time for hydrogen cars being gently launched into the global marketplace. In 2008, Honda launched the FCX Clarity, which was available for lease to customers in Japan and Southern California, although it was relegated to the big carpark in the sky in 2015. Around 20 other hydrogen-powered vehicles were released in either prototype or demo form, including theF-cell hydrogen fuel cell electric vehicle (FCEV—not FCV, as some people label it) from Mercedes-Benz, the HydroGen4 from General Motors, and the Hyundai ix35 FCEV." (Corby, 2021) According to a study from information Trends (via PR Newswire), just over 56,000 hydrogen FCEVs were sold globally by 2022. While the demand is growing, it's still at a basic level." (Parikh, 2023)

The need for speed

Automobile manufacturers, whenever they build an alternative fuel car, feel that they must convince the public that they can produce high speed vehicles. The electric vehicle builders competed with one another to show how fast their electric powered cars would go. Similarly, manufacturers of hydrogen powered cars felt this need for speed. "Momentum is building behind hydrogen as a solution for zero-emission motorsport. But what challenges face those taking the plunge, and how long will these take to iron out?" (Newbold, 2023) A main concern of electric vehicles is battery life. This is a reason why manufacturers are looking for other ways to power vehicles. Hence, the consideration of hydrogen powered vehicles is growing. "Motor racing has always challenged the boundaries of what is possible from a technological perspective and has often been instrumental in driving change in the consumer market." (McEvoy, 2021) Braking regeneration is one example.

In 2004 the BMW H2R set nine world speed records with its streamlined, aluminum space frame chassis and racing suspension vehicle with a top speed of 187.62 mph. This was not a hydrogen fuel cell car. It had an internal combustion V12 engine that was modified to run on hydrogen. In the development of this hydrogen powered record setting race car, BMW engineers and design specialists were supported by three factors. "First, the components featured in BMW's hydrogen production car of the future have now reached a high degree of maturity allowing their unproblematic and straightforward adaptation for the record car. Second, the development specialists were able in the development process to use proven BMW chassis and suspension systems naturally meeting the strictest requirements. And third, consistent, far-reaching use of CAD technology allowed a clearly oriented and time-saving development process." (Hanlon, 2004) "A fundamental consideration is that the combustion properties of hydrogen are quite different from those of gasoline or diesel. While hydrogen burns faster than conventional fuels under normal air pressure, the combustion temperature is slightly lower than in the case of gasoline. A significant advantage of the higher combustion pressure of the hydrogen/air mixture is that the generation of more power from the same amount of energy means a higher degree of efficiency. Before the hydrogen/air mixture is able to flow into the cylinders, the combustion chambers are cooled by air to ensure that the fuel/air mixture is not able to ignite in an undesired, uncontrolled process." (Hanlon, 2004) The sleek bodied racer with the characteristic BMW grill set the records at a racecourse in the south of France. BMW used much of the technology to develop hybrid cars with an eye to the future of hydrogen fuel cell vehicles.

In 2004, a team of engineering students from Ohio State University, Ballard Power Systems and Roush built a streamliner to attempt speed records at Bonneville Salt Flats. The hydrogen fuel cell powered streamliner hit speed trap speeds of 314.958 mph. At the time that was the fastest land speed record for an electric vehicle. Called the Buckeye Bullet, the same group came back and attempted, but failed to beat their own time. After using a lithium-ion powered streamliner the Ohio State team set a new record with the Buckeye Bullet 3. "After three years of battling difficult weather conditions at the Wendover, Utah, Bonneville Salt Flats track. The Ohio State University's Venturi Buckeye Bullet 3 student team and driver Roger Schroer rallied to push their electric streamliner vehicle to a world record two-way average top speed of 341.4 miles per hour (549.4 kilometers per hour) on Monday, Sept. 19, 2016."(Ohio State News) "Venturi Automobiles was founded in 1984 in France as a sports car manufacturer. Purchased in 2000 by Gildo Pallanca Pastor, it focses on innovation in the field of electric vehicles by harnessing the most advanced technological solutions in this area. Venturi expertise extends from urban cars to high performance vehicles."(Ohio-StateNews, 2016)

In 2007, at the Bonneville Salt Flats in Utah, Ford Motor Company became the first manufacturer of a production vehicle, the Ford Fusion, powered by a hydrogen fuel cell car to set a land speed record of 207.297 mph. "The car was designed and built by Ford engineers in collaboration with Ohio State University, Ballard Power Systems and Roush."(Auto Power Girl, 2007) "What we accomplished is nothing short of an industry first,' said Gerhard Schmidt, vice president, Research & Advanced Engineering for Ford Motor Company. 'No other automaker in the world has come close. We are excited to have accomplished something never done. We established this project to advance fuel-cell powered vehicles and to do what has never been done before, and we did it."(Auto Power Girl, 2007) "Hydrogen fuel cells combine hydrogen, (the most common element in the universe) and oxygen to generate energy. Importantly, the only by-product ODF such a union is H2O, which makes it very attractive to those looking for a greener source of power. The Ford Focus 999 project aims to showcase the potential of hydrogen fuel cells to be just as effective, and efficient, as traditional combustion engines. 'Our goal is to make sure that whatever we're building within our research labs is a high-quality and efficient alternative fuel vehicle so that we can have fuel cell vehicles that the average person could be driving in the relatively near future', said Matt Zuchlk, lead engineer of the Fusion project."(Sherer, 2007)

In 2009, at the El Mirage Dry Lake Bed in the Mojave Deseret, the Quantum streamliner, driven by Jessie James and powered with hydrogen set a one run land speed record of 199.7 mph beating the H2R BMW's speed of 186 mph. "JesseJames' record setting run was supported by 24 Quantum hydrogen injectors that fueled an 8-cylinder engine that was optimized to yield 704 HP."(Hardigree, 2009) This, like the H2R were hydrogen powered vehicles, not hydrogen fuel celled vehicles. But they demonstrated the use of hydrogen as a fuel source for vehicles. However, information gathered by these attempts at speed with hydrogen power opened the door for research on hydrogen fuel cell technology. "While completely impractical now, there are reasons to keep research into hydrogen fuels ongoing. A fuel cell vehicle—in theory—is a radically different approach to transportation than that taken by internal combustion engines. As in battery-electric vehicles, a fuel cell vehicle (FCV) is driven by an electric motor or motors. However, battery vehicles use electricity from an external source, while fuel cell vehicles can create their own electricity. "(Zino, 2009)

In 2020, Hyperion announced the hydrogen hypercar called the Hyperion XP-1. This supercar is reportedly able to exceed 200 plus miles per hour and have the ability to go 1,000 miles on a full fuel load. Hyperion has been working on the Hyperion XP-1 for a decade and they expected to have the first car roll off the assembly line in 2022. The Covid-19 virus slowed the production process. The Hyperion XP-1 is a hydrogen fueled car not a hydrogen fuel cell car. The Hyperion is lightweight and expected to go from 0 to 60 MPH in just 2.2 seconds. That's power. Because it is fueled with hydrogen it is environmentally very clean with just water vapor exhaust. Who is Hyperion and what is its mission? "Hyperion is first and foremost an energy company whose mission is d to deliver clean, renewable power to the world.' Angelo Kafantaris, the CEO of Hyperion told me recently.' We've identified hydrogen as the best energy carrier to do that.' Kafantaris said building a supercar is simply the best way to tell that story and called the XP-1 the first chapter of Hyperion.' The other chapters are designed to explain how Hyperion is going to bring hydrogen to the rest of the world." (Blanco, 2020) "Kafantaris says, 'We're an energy company that's building this car to tell a story." (Dyer, 2020) Hyperion also plans to build the hydrogen fuel infrastructure to make hydrogen fueling station available for Hyperion owners. "Besides the Hyperion XP-1, the company which supplies hydrogen propulsion systems to other firms announced the XP-7 Hyper:Fuel power stations using NASAS technology to refuel both FCEVs and BEVs. The stations will also be capable of providing 'utility grid support for emergency and backup power applications." (Pappas, 2022) The Hyperion made its debut at the Los Angeles Auto Show in 2022. Plans are to produce 300 units once production begins.

"The H2-powered Toyota Corolla has completed a 24-hour race at the Fuji International Speedway." (Wallace, 2023) Notice that this Corolla was powered by a liquid hydrogen engine and not a hydrogen fuel-cell engine. Toyota had competed in the Super Taikyhu Series since the third round of the 2021 races, but this was the first time a hydrogen powered Corolla was raced. The car had a hydrogen fueled internal combustion engine. "Toyota is using the race car to learn more about its operation than is available through research." (Wallace, 2023)

"The automotive world isn't just looking to curb carbon emissions for production cars. It's also trying to cut emissions on the racetrack." (Motor Authority, 2024) One of the most significant recognitions of the future of hydrogen cars is the recognition by the 24 Hours of Le Mans rules committee announcement in 2023, that there will be a category for hydrogen and hydrogen fuel cell vehicles starting in 2025. The 2023 Le Mans was actually the 100th year of the event. The first actual race for H cars will likely be in 2026 Le Mans. "Mobility is undergoing a revolution, that of the energy transition and the 24 Hours of Le Mans participate in the search for sustainable technologies. Since 2018, the ACO, with Mission H24 and Green GT, has been promoting hydrogen, a safe and efficient energy carrier. This day is a new crucial step for the creation of the Hydrogen category of the 24 Hours of Le Mans." (Automotive, 2023) The mission H24 competed in at Le Mans in a supporting Le Mans Cup event. President Fillon emphasized that the ACO-led Mission H24 electric-hydrogen prototype should be separately from the Le Mans hydrogen program. "We ae not a manufacturer. The car is just a laboratory to learn what we must do in terms of safety and refueling. We have learned a lot with this car." (Kilbey, 2023)

Since this announcement the ACO has adjusted their time frame for hydrogen racers at Le Mans. ACO president Pierre Fillon citied hurdles as the reason for the postponement. "2026 is not realistic, it's (now) 2027." (Kilbey, 2023) "A new hydrogen category will be introduced in 2026, with cars either using fuel cell technology or combustion engines running on hydrogen, and all hypercars will eventually be hydrogen-powered. The goal is to have all the top category in 2030 with hydrogen, said Fillon." (Balwin, 2023)

Toyota was quick to respond with the introduction of the Toyota GR H2 Racing Concept. Toyota plans to bring hydrogen technology to the 24 Hours of Le Mans in the next three years. "Hydrogen power isn't a brand-new concept to Toyota. Research and development for the first Toyota hydrogen car date back to 1992." (Hurllin, 2023) The GR H2 Concept will also be powered by a hydrogen fueled internal combustion engine. "The sound and torque and dynamics, it's all there, explained company chairman Aklo Toyoda on the eve of last weekend's race. 'Personally, my goal is to achieve carbon neutrality in motorsport without sacrificing performance or excitement. Le Mans is a very special place for Toyota, a place where we not only compete in one of the world's most celebrated races, but a place where we can push the boundaries of technologies. A place where we can realize the future." (Holding, 2023) Hyundai is also known to be looking at options to run a parallel a high bred hydrogen powered LMH.

Students at T U Delft introduced the "FORZE IX," a hydrogen fuel cell prototype with an estimated top speed of 300 kph. The Forze IX was designed to compete in the ICE P and L series in an endurance race at the Zandvoort Circuit in the Netherlands. "We want to keep challenging ourselves, 'says the Forze IX director and team manager, Coen Tonnager. The Froze IX will have a peak power of around 800 bhp. This power is generated by the fuel cells, but we also have an additional supercapacitor energy storage system that can give a huge power boost to the car. There's a lot of performance inside this vehicle." (Barras, 2022)

The Forze IX's, sponsored by Shell, most technical advancement is the supercapacitor. "Sometimes called an 'ultracapacitor' or 'high-energy capacitor'" a supercapacitor, like a battery, stores, and releases electricity. Unlike a battery, which stores its energy in chemicals, supercapacitors store electricity in a static state. Like the static electricity that zaps you when you walk across the living room carpet to turn on a lamp, making it possible to release a bunch of energy in much less time than a conventional battery." (Barras, 2022)

"Alpine is developing a hydrogen fuel-cell racer for Le Mans" (Barlow, 2022) Alpine would target a Garage 56 entry, which allows a grid slot for competitors wanting to test new technologies. Alpine is also considering an assault on the lap record at the Nurburgring using a hydrogen fueled engine.

"Ligier, Bosch join forces on hydrogen car to be revealed at Le Mans. Sportscar constructor Ligier Automotive will reveal a new hydrogen demonstration vehicle at next month's Le Mans 24 Hours after embarking on a joint venture with Bosch Engineering." (Newbold, 2023) The car uses the Ligier JS2 but with a completely different chassis. The Ligier will be called the Ligier JS2 RH2 and will be hydrogen powered but not by hydrogen fuel-cells. Presently it is for demonstration purposes but one version has been driven on test tracks in Germany. "Dr. Johannes-Jorg Ruger, president of Bosch Engineering, said; Hydrogen engines offer a huge potential for high-performance applications, especially in motorsports. By constructing the demonstration vehicle, we illustrate our many years of expertise as an engineering service provider and in particular, our competence in the complex environment of hydrogen." (Automotive, 2023)

"Ligier Automotive President Jacques Nicolet; 'As a manufacturer of racing cars and special vehicles, we must provide the innovations to meet tomorrow's challenges in order to offer motorsport is part of Ligier Automotive's strategy to become a preferred partner of automotive manufacturers for integrating new energies and new technologies." Newbold, 2023) and high-performance vehicles a new path for development. This project

Hydrogen fuel power

The question exists as to why hydrogen power and not the current trend towards lithium-ion battery power? Glenn Liewellyn, Vice president at airbus offered this answer. "The question is how big can we go with batteries, said Glenn Liewellyn, vice president of zero-emissions technology at Airbus, in a briefing, 'We don't believe that it's a today-relevant technology for commercial aircraft and we see hydrogen having more potential." (Ryan, 2020) There is one estimate that by the year 2030, one of every twelve cars in California, Germany, Japan, and South Korea will be hydrogen powered. "Hydrogen fuel cell cars are becoming a popular option for car makers seeking to overcome the restrictions from the limited range and long charging times of battery-based electric vehicles." (Autocar, 2019)

The BMW H2R, Buckeye Bullet #1, and Jesse James hydrogen powered land speed record holder are not hydrogen fuel cell vehicles. Instead, they illustrate the use of hydrogen as a fuel for modified internal combustion engines. The adoption of hydrogen fuel is finding many uses as a zero-emission power source.

In September of 2020, Airbus, a French aviation company, revealed plans to produce zero-emission aircraft fueled by hydrogen. Actually, Airbus introduced three concept aircraft. "Guillaume Faury, the Airbus chief executive, said the 'historic moment for the commercial aviation sector' marks the 'most important transition this industry has ever seen.' 'The concepts we unveil today offer the world a glimpse of our ambition to drive a bold vision for the future of zero-emission flight. I strongly believe that the use of hydrogen—both in synthetic fuels and as a primary power source for commercial aircraft—has the potential to significantly reduce aviation's climate impact,' he said."(Ambrose, 2020) All three of the Airbus concepts would use hydrogen as fuel. One of the three concepts is a turbofan powered craft. "In the turbo fan design, liquid hydrogen will be stored and distributed through tanks located behind the rear pressure bulkhead, while at the same time hydrogen fuel cells will create electric power that compliments the gas turbine. "(Ryan, 2020) When hydrogen is burned it produces water vapor as an emission, making it a clean fuel for planes, trains, and trucks. Faury also indicated the need for the infrastructure at airports for hydrogen fueling stations to make hydrogen powered aviation a reality. If these three concepts are successful it will revolutionize, and industry currently fueled by fossil fuels.

"Fuel cells are inherently more efficient than internal combustion engines (ICE), which must first convert chemical potential energy into heat, and then mechanical work. Fuel cells are also cleaner than traditional fuel sources: the only byproduct is warm air, vapor, and H2)O—water—and its clean enough to drink. But that's once hydrogen is harvested. The hydrogen fuel concept compliments Airbus' overarching goal to reduce fossil fuel consumption in its future fleets." (Cohen, 2020)

Besides aviation, train transportation is another possible area for hydrogen fuel. A European company, Alstrom, has developed a train that harnesses fuel cell technology to power its trains while emitting only steam and water. In 2016, Germany introduced the hydrogen-powered train and in 2018, the train was put into service. The German, Coradia iLint trains had a range of

600 miles about the same as diesel powered trains. The same French company, Alstom, that developed the German hydrogen-powered train is looking at the North American market to replace the diesel-powered trains. In 2015, China put their first fuel cell train into operation and the Chinese government officials are backing hydrogen-power development. Hyundai moves ahead with a multi-billion-dollar plan to build hydrogen powered trucks, trains, and ships. Toyota is building a fleet of fuel cell trucks for the Los Angeles area. Swiss supplier, Stadler, plans to operate a FFLIRT H2 train with the fuel cell and hydrogen tanks mounted on the roof, operating in southern California in 2024.

"The 2010s saw battery-powered electric cars move into the mainstream, led by Elon Musk's Tesla, and as that trend gains steam there are signs the decade ahead will see hydrogen gain commercial viability in transportation, particularly for heavy vehicles like trains and long-haul semis that need a more flexible power train than multi-ton battery packs." (Ohnsman, 2018) " Freight locomotives for long-distance hauling is the most technically challenging, but has the highest societal value in that the diesel volume displacement with hydrogen fuel would add significantly to economies of scale and reduced fuel cost, ' according to a recent report for the Energy Department and Federal Rail Administration prepared by Sandia National Laboratories."(Ohnsman, 2019) "The Canadian Pacific Railway (CP) announces plans to develop what it hopes will be North America's first hydrogen-powered locomotives for freight trains. As part of a pilot project, it will with hydrogen fuel cells and battery technology. It will then test the locomotive to evaluate its use for hauling freight." (Desjardins, 2020)

In 2012, EPA and the National Highway Traffic Safety Administration unveiled the most dramatic overhaul to vehicle fuel economy labels since they were introduced more than 35 years ago. The redesigned label includes how much you'll save or spend on fuel over the next five years, compared to the average new vehicle. Ratings on a vehicle's smog and greenhouse gas emissions. Driving range and charging time for electric vehicles. And a QR Code to access additional information online on your smartphone.

Five Common Fuel Cell Myths

Myth #1: Refueling a hydrogen car is more difficult than a gasoline car.

It's actually a very similar process to fueling from a gas pump. You simply insert the pump into the receptacle in the same way, secure the nozzle and turn on the pump. The whole process takes 5-8 minutes to completely refuel.

Myth #2: Fuel cell cars emit way more water than gasoline cars.

Fuel cell cars produce roughly the same amount of water as gasoline powered cars. Remember the fuel cell exhaust is water vapor which rises into the atmosphere harmlessly.

Myth #3: Hydrogen is difficult and expensive to produce.

Tons of hydrogen are produced every year. Just as every product once the demand occurs and production increases the cost of producing hydrogen for fuel cell vehicles will be reduced. Mining is not required and most likely the reader conducted an electrolysis experiment in school science classes where water is separated into Hydrogen and Oxygen.

Myth #4: Fuel cells are expensive and will never make it mainstream.

History has proven wrong because fuel-cell powered cars and trucks are now in production. Each year new technology is producing better and cheaper fuel cells. Fuel cells are being used for other purposes. Sprint and AT&T are using fuel cell systems as backups to power cellphone towers across the country.

Myth #5: Hydrogen is not safe.

Hydrogen is just as safe as any other transportation fuel. The composite fiber tanks have been rigorously tested, even being fired on by rifles, and proven safe.

Benefits

- Fun to Drive

- Instant torque and smooth consistent power

- High-tech

- Low maintenance

- Zero emissions

- Fast refueling (3–5 minutes)

- High range

- Access to carpool lanes and other incentives

- Attractive lease pricing often bundled with free fuel/maintenance.

(Drive Clean, 2021)

4.

HOW DOES a

HYDROGEN FUEL CELL WORK?

Fuel Cell History

"In 1839, the first fuel cell was conceived by Sir William Robert Grove, a Welsh judge, inventor, and physicist. He mixed hydrogen and oxygen in the presence of an electrolyte and produced electricity and water. The invention, which later became known as a fuel cell, didn't produce enough electricity to be useful."(Bellis, 2019) The term "fuel cell," was coined in 1889. However, there is a dispute as to who gets the credit for coining the term. "In October of 1959, Harry Karl Ihrig, an engineer for the Allis-Chalmers Manufacturing Company, demonstrated a 20-horsepower tractor that was the first vehicle ever produced by a fuel cell." (Bellis, 2019)

How does it work?

Fuel cell electric vehicles (FCEVs) are powered by hydrogen. They are more efficient than conventional internal combustion engine vehicles and produce no tailpipe emissions. They only emit water vapor and warm air. FCEVs use a propulsion system similar to that of electric vehicles, whereas energy stored as hydrogen is converted to electricity by the fuel cell. Unlike the conventional internal combustion engine vehicles, these vehicles emit no harmful emissions.

"A hydrogen fuel-cell vehicle (HFCV for short) uses the same kind of electric motor to turn the wheels that a battery-electric car does. But it's powered not by a large, heavy battery but by a fuel-cell stack in which pure hydrogen ($H2$) passes through a membrane to combine with oxygen ($O2$) from the air, producing the electricity that turns the wheels plus water vapor." (Voelcker, 2022)

"At their most basic, fuel cells combine hydrogen and oxygen through a permeable membrane coated with rare metals, such as platinum. The process creates a flow of current and water vapor as the only exhaust. That energy can be used to power the same sort of electric motor configuration found in battery-electric vehicles—which is why the technology is often referred to as a refillable battery." (Eisenstein, 2023) "The fuel cell is the system's power plant. In it, hydrogen gas drawn from an onboard pressurized tank reacts with a catalyst, typically made of platinum. The process strips the electrons from the hydrogen, freeing them to do their thing—which is to be the electricity that flows through the electric motor to power the car. One fuel cell doesn't produce all that many electrons, so automakers bind scores of flat, rectangular cells together into a fuel-cell stack to get enough juice to power a car or truck. The stack acts much like a battery, releasing electricity in a constant flow to power the vehicle's electric motor and auxiliary electronics. Once the fuel-cell stack does its magic, the vehicle is just like any other electric-drive vehicle on the road running in near silence with loads of acceleration, thanks to the electric motor's hefty torque output."(Edmunds, 2015)

Hydrogen atoms enter the anode. The atoms are stripped of their electrons in the anode. The positively charged protons pass through the membrane to the cathode and the negatively charged electrons are forced through a circuit, generating electricity. After passing through the circuit, the electrons combine with the protons and oxygen from the air to generate the fuel cell's byproducts water and heat.

"By 2030, hydrogen is expected to be capable of moving 10-15 million passenger cars and half a million trucks, Hydrogen fuel cell electric vehicles are an alternative to battery-operated electrics. They offer greater autonomy, faster recharging times and therefore allow recurrent use of the vehicle, something for which current batteries are more limited." (Frey, 2023)

Fuel Cell electric technology explained

"A hydrogen car is a type of electrified vehicle. So, what we do is we move the wheels with electric motors, but the power that drives these motors comes from the fuel cell. And what a fuel cell is is it's basically a stack of individual cells—there might be about 300 of them in a stack—and in one side we bring hydrogen, and on the other side, we bring in air with oxygen. And there's a reaction that directly converts those two into water and makes power, electric power. That's what we generate, the power that moves the traction drive." (WBUR, 2017)

- **Motor:** Motor driven by electricity generated by the fuel stack and supplied by battery.

- **Power Control Unit:** A mechanism to optimally control both fuel stack output under various operational conditions and drive battery charging and discharging.

- **Fuel Cell Stack:** Toyota's first mass-production fuel cell, featuring a compact size and world top level output density.

- **Fuel Cell Boost Converter:** A compact high efficiency, high-capacity converter newly developed to boost fuel cell stack voltage to 650V. A boost converter is used to obtain an output with higher voltage than the input.

- **High Pressure Hydrogen Tanks:** Tank storing hydrogen as fuel. The nominal working pressure is a high-pressure level of 70MPa(700bar). The compact, light-weight tanks feature the world's top level tank storage density.

- **Battery:** A nickel-metal hydride battery which stores energy recovered from deceleration and assists fuel cell stack output during acceleration.

(Frey, 2023)

5.

HOW is a

HYDROGEN FUEL CELL USED?

"HYDROGEN FUEL CELLS USE HYDROGEN AS a fuel in an electrochemical process that combines hydrogen and oxygen to produce electrical energy and water. The reverse process of electrolysis, which produces 'green' hydrogen and oxygen from water, can use a range of renewable energy resources (wind, wave, solar) to produce hydrogen as a fuel for renewable power generation. There is also growing interest in hydrogen power as a uniquely clean energy source that can produce heat and whose only by products are water." (TWI, 2022) Hydrogen is a clean and flexible energy source and therefore is being considered for other energy needs like home heating.

"Where local conditions allow, the availability of hydrogen through local generation and storage could prove to be an alternative to diesel-based power and heating in remote areas. Not only will this reduce the need to transport fuels but will also improve the lives of those living in distant regions by offering a non-polluting fuel obtained from a readily available natural resource." (TWI, 2022) "Hydrogen powered vehicles are just one example, but it could be used in smaller applications such as domestic products as well as larger scale heating systems." (TWI, 2022)

"The European Union (EU) is pushing a hydrogen program, which will also be aimed at aviation and heavy industry." (Winton, 2020) "General Motors is finding new markets for its hydrogen fuel cell systems, announcing that it will work with another company to build electricity generators, electric vehicle charging stations and power generators for military camps. The emission free generators will be designed to power large commercial buildings in the event of a power outage, but the company says it's possible that smaller ones could someday be marked for home use." (Associated Press, 2022)

The generators would be very quiet. These generators could charge police and emergency vehicles in the case of a power outage. Individual residents would be able to use a hydrogen fuel cell generator for home use in a power outage. Some warehouses already have fuel cell powered forklifts in operation. Because fuel cells can be grid-independent, they're also an attractive option for critical load functions such as data centers, telecommunications tower, hospital, emergency response systems, and even military applications for national defense. Currently there are drones which use hydrogen fuel cell power. Research and development is being directed at underwater drones.

WHA-International has produced a list of ten applications of hydrogen fuel cells.

1. **Warehouse logistics** – Hydrogen fuel cells are powering trucks, forklifts, pellet jacks, and more.

2. **Global Distribution** – fuel cells have both the range and power required for long haul trucks and local distribution.

3. **Buses** – Buses that are hydrogen fuel cell powered are currently in use world-wide.

4. **Trains** – Hydrogen powered trains are now in use in Japan, South Korea and many European countries.

5. **Personal Vehicles** – Automobile manufacturers are currently building HFCEVs.

6. **Planes** – As of 2023, several manufacturers are developing hydrogen fuel cell powered airplanes.

7. **Backup Power Generation** – Businesses including hospitals and data center are using stationary hydrogen fuel cell power plants as a backup in the event the electricity goes out.

8. **Mobile Power Generation** – some of the earliest applications were developed by NASA for rocket and space shuttle systems.

9. **Unmanned Ariel Vehicles (UAVs)** – Both military and private business systems are using drones (UAVs) powered by fuel cells.

10. **Boats and Submarines** – In some German submarines Hydrogen fuel cell s offer an alternative to nuclear power with long range silent cruising.

(WHA-International, 2023)

Ballard industries has developed a stationary power system that is designed to generate clean energy from hydrogen. The generator features electrical output that can be either AC or DC with a range of voltages. This system produces high electrical efficiency with average fuel consumption. The system also produces zero-emissions with only pure water discharge. The key advantages of the Ballard system are low operating cost, flexible solution, high reliability and safe operation.

English engineer Francis Thomas Bacon developed the world's first hydrogen/oxygen fuel cell in 1932. "Bacon's fuel cell was such a success that it has been used by the space industry to power satellites and rockets for space exploration programs, including Apollo 11, since the 1960s. As the story goes, then President Richard Nixon famously said, 'Without you Tom, we wouldn't have gotten to the moon." (Innovation, 2020) This same technology is now being studied for possible aviation applications. The possibility exists that someday aircraft will be powered by hydrogen fuel cell technology.

Here are ten possible applications of hydrogen fuel cells.

1. Warehouse logistics

2. Global distribution

3. Buses

4. Trains

5. Personal vehicles

6. Planes

7. Backup power generation

8. Mobile power generation

9. Unmanned arial vehicles (UAVs)

10. Boats and submarines

"The European Union (EU) is pushing a hydrogen program, which will also be aimed at aviation and heavy industry." (Winton, 2020)

6.

WHAT are the ADVANTAGES of a HYDROGEN FUEL CELL?

C HINA HOSTED THE WINTER OLYMPICS IN 2022. Hydrogen fuel cell vehicles were showcased. Some were concerned about the hydrogen fuel cell vehicles to perform. Autotrader in February 2022 offered this rebuttal. "However, the Winter Olympics 2022 in China have showcased that hydrogen-powered vehicles are a viable alternative to their electric counterparts, with over 800 hydrogen buses and thousands of hydrogen cars serving the Olympics and the Zhangjiakou area proudly, so why haven't they taken off in the UK?" (Lex, 2022)

1. Zero emissions from production to fueling

"The draw of hydrogen-powered cars is that they allow owners to maintain the habits they've built with petrol and diesel cars." (Carexpert)" Here's what you need to know about these cars and the safety of this unfamiliar fuel. The vehicles don't need gasoline. They aren't tied to an electric plug. They produce zero emissions from the tailpipe and can deliver 300 miles or more per tank of fuel. They can be refilled as fast—or faster—than a conventional car with a 15-gallon gas tank. After extensive testing, researchers say they are as safe to drive as gasoline cars."(Edmunds, 2015)

"Fuel cell technology advancements have improved the viability of hydrogen-powered vehicles, with the weight of the fuel cells reducing and the efficiencies improving. "(TWI, 2022)

2. Longer range with a fill up

Hydrogen fuel cell vehicles FCEVs are well suited for long range and heavy payload vehicles. FCVEs are also well suited for city buses and trains. FCEVs produce 20–30 percent less emissions than conventional gasoline fueled cars. In 2021, a Toyota Mirai with professional drivers behind the wheel, set a Guinness World Record for distance, with an 845-mile run. A record that far out distances the range of current electric vehicles in 2022.

That record more than doubles the Environmental Protection Agency's official range rating.

3. Greater percent of efficiency

Standard combustion engine cars run on about 20% efficiency. In contrast a hydrogen fuel cell vehicle would operate at 40% to 60% efficiency. The U.S. Department of Energy estimates that maximum efficiency of a gas combustion car would be around 58% whereas the hydrogen fuel cell could max out at 85–90% efficiency. Hydrogen fuel cells can produce more power and energy efficient vehicles than those with fossil fuels.

Hydrogen fuel celled vehicles operate at greater efficiency than gas combustion vehicles. Hydrogen fuel cells are more powerful and energy efficient than fossil fuel vehicles. Hydrogen fuel cell cars have a greater driving range. It also takes less time to refuel a hydrogen fuel cell car.

Key benefits include:

- Unlike gasoline- and diesel-powered vehicles, powered by hydrogen fuel cells do not produce air-pollutants.

- It reduces dependence on petroleum imports as hydrogen can be domestically produced from various sources.

- Fuel cell vehicles powered by hydrogen do not produce greenhouse gas emissions. (Walker, 2020)

- Hydrogen tanks allow for the hydrogen gas to be stored for later use. Refueling Hydrogen vehicles can be faster than gas electric battery powered vehicles.

- Hydrogen fuel cell vehicles are lighter than fossil fuel vehicles making them more efficient.

- Compared to electric cars, whose batteries deplete faster in cold temperatures, hydrogen cars are better suited for wintry climates.

4. Lack of the need for rare metals

What of the greatest advantages of the hydrogen fuel cell is the lack of the need for metals needed for electric car batteries like cobalt, lithium-carbonate, and copper. Green Car Congress points out the dramatic truth about the use of these metals in electric car batteries. "The metal resource needed to make all card and vans electric by 2050 and all sales to be purely battery-electric by 2035. To replace all UK-based vehicles today with electric vehicles (not including the LGV and HGV fleets), assuming they use the most resource-frugal next-generation NMC811 batteries, would take 207,900 tonnes cobalt, 264,600 tonnes of lithium-carbonate (LCE), at least 7,200 tonnes of neodymium and dysprosium, in addition to 2,363,500 tonnes copper. This represents just under two times the total annual world cobalt production, nearly the entire world production of neodymium, three quarters the world's lithium production and at least half of the world's copper production during 2018. Even assuring the annual supply of electric vehicles only, from 2035 as pledged, will require the UK to annually import the equivalent of the entire annual cobalt needs of European industry." (Green Car Congress, 2021)

5. No dependance on China

Hydrogen is abundant and does not require mining. Because Hydrogen does not depend on mining, China which controls the lithium mining market cannot control hydrogen fuel cell production in the United States

6. Fast charging times

Hydrogen fuel cells are refueled from a pump much like regular gas. Hydrogen fuel cells can be refueled in 5–8 minutes.

11. No noise pollution

12. Hydrogen fuel cell vehicles are ideally suited for rural areas

13. Hydrogen fuel cell power is versatile in its many uses

14. Renewable and readily available

Keith Miller writing in his blog gives these nine advantages of hydrogen fuel cells.

1. It offers an effective method of energy storage.

2. This technology offers a high level of energy efficiency.

3. The emissions from a hydrogen fuel cell are virtually zero.

4. Vehicles using a hydrogen fuel cell achieve a better fuel economy rating.

5. We receive a greater level of consistency with hydrogen fuel cells.

6. It is possible to create hydrogen fuel cells with a neutral emissions cost

7. A hydrogen fuel cell provides us with energy flexibility.

8. Hydrogen fuel cells are a safe technology for us to use in virtually any situation.

9. You can reduce the risk of chemical exposure by using hydrogen fuel cells.

(Miller, 2021)

7.

WHAT are the DISADVANTAGES
of a HYDROGEN FUEL CELL?

"HYDROGEN POWER HAS BEEN AROUND for decades, used in past NASA spacecraft, but costly materials, including precious metals in fuel cell membrane that help generate electricity and carbon fiber fuel tanks, as well as long-term durability issues limited its appeal for transportation. What's more, industrial production of hydrogen from natural gas, with by-product carbon emissions, undercut its appeal as a green fuel. But aggressive R&D over the past decade has chipped away at costs, with synthetic alternatives replacing metals like platinum in fuel cells, while durability and performance in extreme heat and cold continue to improve. "(Ohnsman, 2021)

"Another argument that's often made against hydrogen cars is that they're less efficient than electric cars. That's because the hydrogen required to power the vehicle isn't naturally occurring, meaning it has to be extracted then compressed for use in fuel tanks." (Lex, 2022) "BEVs are considerably more efficient than FCVs, when you take into account the whole series of steps between power generation and propulsion." (Morris, 2020) Hydrogen still has its main strengths in that it has lightness and quick refueling. "So, for

industrial vehicles, hydrogen seems a viable option, despite the inefficiency."
(M0rris, 2020) "Hydrogen-powered vehicles require more manufacturing
materials than electric cars, so although there are no harmful emissions from
exhaust, hydrogen cars have an increased impact on the environment during
their life cycle." (Lex, 2022)

"Even though the only direct waste product of hydrogen fuel is water,
obtaining this form of power is not necessarily squeaky clean. The cheapest
and most common method at present uses natural gas and high-temperature
steam." (Hirschlag, 2020) Storage is also a major problem of using hydrogen power.

When hydrogen is mentioned as a fuel, the first thing that comes to
mind is the Hindenburg explosion of 1937. Regarding the disadvantages
of hydrogen fuel celled cars, Fred Lambert, writing in Electrek offered this
thought. "My concerns were always more about the efficiency of the power
train and the entire supply chain versus the simplicity and efficiency of
electricity and battery-electric vehicles, but we also can't avoid the safety
concerns of carrying and storing something like hydrogen." (Lamber, 2019)
"HFCVs are widely considered as safe as any other car, since the high-pressure tanks are designed to survive even highest-speed crashes without leaking
or breaching," (Voelcker, 20222)

Refueling is a problem because there isn't sufficient infrastructure
of hydrogen fuel pumps. The lack of an infrastructure of fuel cell refueling
station is the biggest drawback for hydrogen fuel cell cars. In 2022, it was
estimated that there were only about 50 fuel cell station, all of which were in
California in the United States. Axel Rucker, program manager for hydrogen
fuel cell cars for BMW had this to say about the infrastructure issue. "We have
a chicken-and-egg problem with hydrogen propulsion. As long as the network of hydrogen filling stations is so thin, the low demand from customers
will not enable profitable series production of fuel cell cars. And as long as
there are hardly any hydrogen cars on the roads, operators will be reluctant
to expand their refueling network," (Automotive, 2022) One concern is the

possibility that the fuel pumps might explode. The hydrogen tanks have been known to crack, causing gas leaks. Hydrogen gas is highly explosive. However, modern technology is overcoming this disadvantage with better design for hydrogen fuel pumps. "So why haven't hydrogen fuel cells vehicles (FCVs) taken off? June 2019 could be the month that scrawled the writing on the wall. No sooner had a chemical plant producing hydrogen in Santa Clara exploded, leaving FCV users in California short of fuel, but just a few days later a refueling station in Sandrivka, Norway also went up in flames. This really brought home the truth that hydrogen can be dangerously explosive gas—as if we didn't know it already." (Morris, 2020) The lack of infrastructure is a major drawback, however, Peugeot is working to overcome this problem. "Thanks to government plans, the number of hydrogen stations in Europe is constantly increasing and Peugeot is working directly with energy suppliers to propose package deals." (Greencar, 2021)

"Among the many, many challenges facing hydrogen as a fuel is where it comes from and how to store it." (Zino, 2009) "Volkswagen concluded: Everything speaks in favor of the battery, and practically nothing speaks in favor of hydrogen." (Berman, 2020)

As pointed out hydrogen fuel cells are not without some disadvantages.

- Hydrogen fuel cell vehicles are currently more expensive than conventional vehicles and hybrids.

- Fuel cell vehicles are not as durable as internal combustion engines in terms of temperature and humidity ranges.

- The availability of hydrogen is limited to certain locations and hydrogen is quite expensive to produce as well.

- The system used for delivering gasoline from refineries to gasoline stations cannot be used for hydrogen." (Walker, 2020)

To overcome the disadvantage of cost, significant cost reductions have already been achieved in the cost of refueling stations, and fuel cell stack production. The cost of hydrogen has also been reduced.

1. Cost of production.

Research and development is a major cost in making hydrogen fuel cell technology viable in the use of production electric vehicles. The cost of raw materials adds to this disadvantage. These costs need to be reduced to make hydrogen fuel cells viable. Therefore, the will to invest time and money must be decided. There are some drawbacks yet to be overcome. Engineers have been working to lower costs by reducing the amount of rare metals, such as platinum, needed in a fuel-cell 'stack' while increasing energy output." (Eisenstein, 2021) Platinum represents one 0of the largest cost components of a direct hydrogen fueled polymer electro

Lyte membrane fuel cell, so there is emphasis on approaches that will increase activity and utilization and reduce the content of current platinum group metal (PGM) and PGM-alloy catalysts, as well as PGM-free catalyst approaches for long-term applications.

2. Infrastructure

Currently there are only a few refueling stations in Southern California. For hydrogen fuel cell vehicles to become marketable nationwide fueling stations will need to be built. "All fuels are flammable and must be handled carefully. However, in many ways, hydrogen is safer to use than conventional fossil fuels. If a leak occurs, lighter-than-air hydrogen gas rises up and disperses rapidly. This non-toxic gas is also safe to breathe." (OCTA, 2024) One proposed design for hydrogen fueled trains has the hydrogen storage tanks on the top so that leaks will dissipate into the atmosphere.

3. Regulatory issues

Advancement in research and development is rapid and regulatory issues are behind. In time the field will require regulations. Will those regulations stifle development. With other forces, fossil fuel cars and electric battery-operated cars apply political pressure to regulate hydrogen fuel cell vehicles. Without clear regulatory frameworks manufacturers will struggle to reach financial investment decisions. Everything is covered by codes and regulations. Development must meet national NFPA and IFC codes.

Many of the hydrogen safety codes and standards today are based on practices from the chemical and aerospace industries. DOE is coordinating the efforts of codes and standards organizations to develop more robust codes and standards that ensure the safe use of hydrogen for transportation and stationary applications. DOE has set ultimate targets for fuel cell system lifetime under realistic operating conditions at 8,000 hours for light-duty vehicles, 30,000 hours for heavy-duty trucks, and 80,000 hours for distributed power systems.

4. Hydrogen storage

"Hydrogen isn't just used as a fuel; it's also used as storage. Electrical batteries are one solution to the problem of intermittent energy supply, but our current battery system faces difficulties in storing the energy required. Operators need a large amount of stored clean energy during non-peak production hours to support a modern electrical grid and the U.S. currently doesn't have the battery capacity to store and use the clean electricity we use during the day to power our homes and cities during the night. One answer to this problem is hydrogen as a storage device. Hydrogen allows vast quantities of clean energy to be stored for long durations for use in peak demand and seasonal energy balancing." (FCHEA, 2023)

"Hydrogen also offers the potential for seasonal energy storage., in addition to short-term load balancing. As the world increasingly transitions to renewable energy, winter months will strain electrical grids as less sunlight will reach the solar farms." (FCHEA. 2023)

Mitsubishi Power has identified two projects for hydrogen storage. One was to store hydrogen for long time in underground in salt domes. The second was cryogenic hydrogen storage. "The Advanced Clean Energy Storage project will produce, store, and transport green hydrogen utility scale for power generation, transportation, and industrial applications in the western U.S. " (Lawson, 2021)

5. Performance and Durability

Fuel cell efficiency and performance must be developed through research of innovative materials and integrated strategies.

6. Highly Flammable

Hydrogen is highly flammable as a fuel cell. Recent development has produced highly resistant fuel cell storage tanks. General Motors has had a fleet of hydrogen fuel cell vehicles running since 2007 and they are very safe. The following comes from a General Motors release. "The way we store it onboard the vehicle is a tank—it's a pressurized tank. And these tanks are carbon-fiber wrapped. Actually, we took a display of one of these tanks to a show, and we went through 38 sawzall blades trying to cut this tank open so you could get a look inside. So they are very strong systems, and they're designed to be that way. And then we certify that the vehicles are going to be safe based on our crash tests and the simulation work that we do." (WBUR, 2017) Hyperion CEO Angelo Kafantaris says that Hyperion has solved the safety issue for their Hyperion XP-1 hypercar. "You can throw our tanks off a building or shoot them with a high-powered rifle. They won't rupture." Dyer, 2020)

"The jury is still out on whether hydrogen will ultimately be our environmental savior, replacing the fossil fuels responsible for global warming and various nagging forms of pollution. Two main hurdles stand in the way of mass production and widespread consumer adoption of hydrogen "fuel-cell" vehicles: the still high cost of producing fuel cells; and the lack of a hydrogen refueling network." (West, 2019)

Keith Miller writing in his blog lists these nine disadvantages of hydrogen fuel cells.

1. Hydrogen fuel cells do not work in every situation as of yet.

2. You must regulate the temperature of a hydrogen fuel cell to maximize its use.

3. There is still some risks to the environment to consider with hydrogen fuel cells.

4. The cost to store hydrogen is expensive enough that it is prohibitive for most people.

5. There are transportation losses to consider with hydrogen as well.

6. It costs more to transport hydrogen than it does most other fuels.

7. This technology is not widely available right now.

8. You will pay a premium price to purchase equipment with hydrogen fuel cells.

9. It is not currently a complete renewable energy resource.

(Miller, 2021)

Elon Musk, founder of Tesla Motors has been quoted as saying that the battle for green personal transport is over and battery electric cars have won. (Morris, 2020)

8.

WHAT HYDROGEN FUEL CELL POWERED VEHICLES are in use or BEING DESIGNED?

B ECAUSE THE DEVELOPMENT OF HYDROGEN fuel cell powered vehicles is increasing rapidly, the information in this section is based on research in 2024. Many new developments could have taken by the time of publication.

"Hydrogen fuel cells power electric motors with water as the only emission. Many companies are betting that hydrogen will be the dominant technology powering larger vehicles, from buses to trains, because it offers higher energy density compared to lithium-ion batteries." (The Guardian, 2020) Technology for hydrogen fuel cells will need to be improved before they can become a replacement for fossil fuel engines. As the automotive industry is slowly warming up to Evs, it's easy to forget that at least three big automakers still haven't given up on hydrogen: BMW, Toyota and Hyundai." (Ramey, 2020) "Perhaps by 2030 hydrogen cars will be like Evs are today: Representing 2% to 3% of all new-car sales, amid a sea of Evs dominating new-car sales, and older gas and diesel-engine vehicles." (Ramey, 2020)

"A hydrogen vehicle is a vehicle that uses hydrogen for motive power. Hydrogen vehicles include hydro-based space rockets, as well as automobiles and other transport vehicles. The power plants of such vehicles convert the chemical energy of hydrogen to mechanical energy either by burning hydrogen in an internal combustion engine or more commonly, by reacting hydrogen with oxygen in a fuel cell to run electric motors. Widespread use of hydrogen for fueling transportation is a key element of a proposed hydrogen economy." (Smartcharger, 2020)

Hyperion

"The Hyperion XP-1 is a futuristic fuel cell vehicle designed by Hyperion, a startup based in California. Hyperion was unveiled in 2020. The XP-1 made its debut in 2022. It boasts a range of up to 1,016 miles per tank, a refuel time of under five minutes, and a combined power output of over 2,000 horsepower. Its lightweight design, made possible by a carbon titanium monocoque structure and lack of a substantial battery pack, allows it to accelerate from zero to sixty mph in only 2.2 seconds. Hyperion has not yet set a release date or pricing for the XP-1." (Williams, 2022) The Hyperion XP-1 is said to be ten years in the making. The Hyperion was announced in 2020 at the Los Angeles Auto Show. "The entire look is unconventional, with atypical proportions, wild intakes, and shapes jutting outward from the bodywork, including the solar panel-covered adjustable blades." (Williams, 2022) The power train has supercapacitors instead of batteries, a three-speed transmission, and four axial-flux electric motors that drive all four wheels. "On board are a fuel cell system and carbon-fibre hydrogen tanks as well as supercapacitors instead of batteries." (Randall, 2022) "The XP-1 Hypercell fuel cell module, for example, is said to use a 'highly efficient fuel cell air space architecture' that uses 'membranes coated with a industry-best catalyst capable of reducing rare earth metals and extending service life,' Performance is said to be a factor of 2 higher, as is efficiency—durability is said to be as much as three times better. However, the baseline on which this improvement is based is not mentioned." (Randall, 2022)

Hyperion had planned to put the XP-1 into production in the year 2022. Originally 300 units were planned.

NamX HUV

NamX is out to revolutionize the driving experience. "Hydrogen car start-up NamX says it has made a 'strategic move' to use an H2 V8 internal combustion engine (ICE) for its first vehicle. "The HUV is part of a large-scale technological project to bring together mobility and environmental preservation through the use of green hydrogen." (Autovista, 2022) NamX plans to introduce two models, an entry level -elvel-rear wheel drive version and a four-wheel drive version. Having previously been marketing its 'Hydrogen Utility Vehicle' (HUV) as a fuel-cell electric model." (Collins, 2023) The name NamX comes from New Automotive and Mobility Exploration. The NamX is designed in cooperation with Pininfarina. One of the innovations of the NamX is its dual fuel tanks. "The NamX car benefits from a double hydrogen tank. In addition to the fixed tank, a removable tank made up of 6 hydrogen capsules allows our consumers to easily obtain hydrogen supplies." (NamX, 2023) Pininfarina gets credit for designing the NamX double tank with the installation of the 6 capsules in the rear of the car. "If you can't find a hydrogen station, NamX will send cartridges out to you at home, or wherever else you need them. And the company says you'll be able to use them to power other devices as well, in time." (Blain, 2022) "Each cartridge, when full holds enough hydrogen to give you about 150 km (93 miles) of driving range." (Blain, 2022)

"For five years, the NamX teams have been working on the premium hydrogen car concept. Starting from nothing we have developed a project based on our convictions, our values, but also our passion. We like to keep moving and we know that the automotive industry cannot continue to evolve in the same way as it did a few decades ago. We are facing new environmental challenges that no longer allow us to simply design new thermal cars. From the outset, we wanted to facilitate mobility while making it greener. To achieve this, we have surrounded ourselves with great names in the automotive field, such as Pininfarina." (Autocar, 2022)

Nikola

Nikola hopes to compete with Tesla. Nikola targeted 2018 production of 25 trucks and expand production in 2021 to 400 trucks. "Fuel cell electric cars have been achingly slow to catch on in the US, but activity in the area has begun to tick up overseas. Fuel cells have also begun to attract attention from truck makers including Nikola, which is apparently back on track to launch fuel cell trucks in the US." (Casey, 2023)

A new name set to enter the pickup truck market is Nikola Badger. The Badger has a rugged look, like the Ford F150 Raptor except it has LED lighting that gives the truck that digital stare and helps set it apart from other pickup trucks. The Badger has only been shown as a four-door pickup so far, but other body styles and options are on the horizon. The Badger is about the same height and length of the Ford F 150, but it is slightly wider. The Badger will have two power trains, one which will have four electric motors, one behind each wheel, which will draw power from a giant lithium-ion battery under the passenger compartment, with a driving range of 300 miles. A second version with a hydrogen-electric system may have a range of 600 miles. Both versions are GM designs. The estimated sales price is between $60,000 to $80,000 depending on trim and options. Some buyers may be eligible for the $7500 tax credit from the federal government. Nikola announced that it will design the truck in house but will out-source production to a manufacturer. Current speculation says that General Motors will be the cooperating partner. GM would also be able to appoint one member to the Nikola board. That speculation has been confirmed by General Motors. Following an 11% investment share by Nikola in General Motors. "This strategic partnership with Nikola, an industry leading disrupter, continues that broader deployment of General Motors' all-new Ultium battery and Hydrotec fuel cell systems. We are growing our presence in multiple high-volume EV segments while building scale to lower battery and fuel cell costs and increase e profitability," said General Motors chairperson and CEO Mary Barra in a statement."(Gitlin, 2020) The deal extends beyond the Badger pickup, which is tentatively scheduled to go into production in 2022. "Nikola will use GM technology throughout its program, including GM's new lithium-ion battery platform. Nikola will use GM's Hydrotec fuel cells in its class 7+ and 8 trucks." (Gitlin, 2020)

Nikola was rocked by the sudden resignation of its founder, Trevor Milton, in 2020. The stock market suffered major losses. In response to the resignation in 2020, Iveco said they would continue to build vehicles in Europe, Republic Service asserted that they were looking to develop waste-collection vehicles for Phoenix. In addiction, General Motors planned to continue working with Nikola to build the HFC Badger pick-up trucks. In 2022, Trevor Milton was convicted of wire and securities fraud. In the first quarter of 2023, Nikola posted large losses bringing the stock value down to under a dollar per share. As a result of these first quarter losses in 2023, Nikola paused production of its vehicles.

Navistar

"Navistar collaborates with General Motors and OneH2 to launch hydrogen truck ecosystem." (Navistar, 2021) "Hydrogen fuel cells offer great promise for heavy duty trucks in applications requiring a higher density of energy, fast refueling and additional range, said Prsio Lisboa, Navistar president and CEO, 'We are excited to provide customers with added flexibility through a new hydrogen truck ecosystem that combines our vehicles with the hydrogen fuel cell technology of General Motors and the modular, mobile and scalable hydrogen production and fueling capabilities of OneH2. And we are pleased that our valued customer, J B Hunt has committed to utilize the solution on dedicated routes and to share key learnings." (Navistar, 2021) OneH2 makes hydrogen fueling stations and modular, mobile and scalable hydrogen production. "Navistar plans to make its first production model International RH Series fuel cell electric vehicle (FCEV) commercially available in model year 2024." (Navistar, 2021) OneH2 would supply hydrogen fueling solutions for Navistar which includes hydrogen production, storage, delivery and safety. Navistar believes its semitruck will have a range of 500 miles and could be refueled in 15 minutes.

Grove

Another new car manufacturer to enter the hydrogen fuel cell market was Grove Hydrogen Automotive, a Chinese auto manufacturer. Grove was born in 2016, became registered in 2018. "Grove is a Global car company aiming to offer a truly clean Automotive experience from Manufacturing to the enjoyment of the car."(Chen, 2019) "Grove Hydrogen was established by the Chinese Institute of Geosciences and Environment, which currently manufactures and distributes hydrogen extracted from industrial waste. The institute says it is working with large Chinese cities to install and expand hydrogen charging infrastructure in the coming years.."(Autocar Pro News, 2019) Grove has made a total commitment to produce only hydrogen fuel cell vehicles. Grove claims to be the world's first hydrogen fuel cell only mass production company. The company is based in Wuhan, China, a city internationally known for another reason in 2020. Grove has entered into an agreement with designer Pininfarina to design its cars. The design center will be in Barcelona, Spain. Grove brought concept vehicles to the Shanghai Auto show including a 4 door sedan and a supercar sports car. Grove intends to produce a sedan, SUV and the supercar, all with the influence of designer Pininfarina. The four-door concept will have a bold grill and a highly sculpted rear diffuser and will feature frameless doors and cameras in place of mirrors. The four door SUV will be called the Obsidian and will have an estimated range of 625 miles. . The factory will be located in Chongqing, China, with sales starting in2019 and full volume production in 2020. Sales will concentrate on China with future expansion into Australia and New Zealand.

Grove has issued the following statement of philosophy: "Grove creates cars from the Earth and for the Earth, we respect the energy and materials the planet has given us. Born from the desire to create a car that is truly clean, from the earth it originates to the roads upon which it allows our passengers to roam, Grove has searched the earth to find the greatest scientific advances to achieve this Goal."(Grove Hydrogen Automotive, 2020) Grove says its cars will be environmentally friendly, have the advantage of clean energy, cost less on the road, require less hydrogen-fueling time and support a longer journey without refueling, compared to traditional vehicles. Grove expanded its philosophy with the following statement: "Not content with making a cleaner more efficient car at Grove, we focus aggressively on lowering our production driven emissions and also the use of precious resources on our planet. Our production process avoids the heavy welding and painting procedures of traditional cars, massively reducing the environmental impact in the manufacture of our cars."(Grove Hydrogen Automotive, 2019) "Grove Hydrogen Automotive Company Limited celebrated its first cars produced from its Nan'an District Chongqing production factory in Southwest China, April 13, 2019."(Chen, 2019)

The highlight of Grove's Shanghai Auto show exhibit was its supercar exhibit. This Pininfarina designed H2 Speeds, supercar followed the current state of the art supercar racers. Very aerodynamic. "The 503-HP H2 Speeds will use the chassis of a LeMans Prototype 2 racing model as a base and rely on a pair of electric motors that draw their power from hydrogen fuel cells for speed."(Bleier, 2019) The estimated top speed is 186 MPH and a zero to 60 time of 3.4 seconds. The car has been track tested by GreenGT, a Franco-Swiss company known for developing clean propulsion systems. "Twelve of these hydrogen race cars will be produced, powered by four electric motors that are good for a total output of 653 horsepower. Pinifarina's partner in the project is GreenGT, a Franco-Swiss company that designs and develops sustainable propulsion systems. Some of the components used in the production H2 Speed from Green GT include the 250kW fuel cell and torque vectoring system."(Smartcharger, 2020) The H2s supercar will only have a production run of 12 track only supercars . The cars are now market ready with a price tag of approximately 2.5 million dollars. There already is a list of prospective buyers.

Riversimple

A small company, a small car, in a small country trying to make a dent in the hydrogen fuel cell vehicle market. In 2016, Riversimple, a Welsh company, produced a small two seater, the Rasa, for the fuel cell market. "Hydrogen car start-up Riversimple this year completed a real-world demonstration of its Rasa Beta fuel-cell electric vehicle (FCEV) as part of a two-year publicity funded project to design a local decarbonized smart energy system in southwest Wales." (Collins, 2022) "The Welsh firm and Siemens UK have signed a memorandum of understanding that will allow Riversimple to use Siemens' resource to bring the Rasa to volume production the company said in a press release." (Riversimple, 2022) King Charles III, of the United Kingdom, actually took the Rasa out for a test drive.

"We set out with a blank slate, to design a car for the world in which we now live, shaped by the best technology available to us and answerable to our most pressing concerns. We set out to build a local car that will take people on their local journeys, in the way that they wish to travel, at a cost that is affordable—and without leaving a heavy footprint of air pollution and environmental degradation. To do this we selected a still evolving, but incredibly promising and safe technology. This is hydrogen fuel cell technology." (Spowers, 2022)

The Rasa has been in existence since 2001 and has been working on the fuel-cell concept since 2009. It is lightweight, aerodynamic with winged doors for entry, and covered back wheels. The Rasa is estimated to have a range of 300 miles but actually it may be more like 116 miles. The body is a combination of CFRP (carbon-fibre reinforced plastic and GFRP (Glass-fibre reinforced plastic). It is powered by four radial, flux-wheel, hub-mounted motors. The energy is supplied by a hydrogen fuel-cell. The current Rasa was designed by Chris Reitz.

As of the time of this writing the Rasa has not been a reality. There is only a waiting list.

Alpine A4810 Concept.

The Alpine A4810 is a French born hypercar. "This hydrogen-powered dream machine blends sleek, F1-inspired design with cutting-edge technology, blurring the lines between concept and reality. Born from the minds of design students and nurtured by Alpine's F1 expertise, the A4810 is more than just a pretty face." (Tariq, 2024) "The Alpine A4810 Concept offers a glimpse into the future with next-generation technology, while still showing classic characteristics from the Alpine brand that was founded back in 1955, this is an innovative concept car that has been created for a sustainable future, a car that boasts the traditional Berlinetta shape, and while the engine and fuel tanks might be built similar to a current hypercar, the subtraction process is proof of some strong innovation. The styling comes with a serious influence from Formula One models, and the combination of intakes, vents, and body panels create an overall impression of lightness, even with a massive glass panel almost stretching from the front straight to the back of this concept, this car looks like a LeMans racecar for the road. "(Meyers, 2024) The A4810 signals a commitment to hydrogen power, paves the way for eco-friendly performance cars, and is inspired by Mont Blanc. The number 4810 is the exact height of Mount Blanc. hAlpine is looking ahead to producing the car in limited numbers. I should note that the Alpine is a hydrogen fueled car not a hydrogen fuel cell fueled car.

Audi

In 2016, Audi introduced a hydrogen fuel cell concept called the Audi h-tron. The h-tron Quattro concept was a sporty SUV that used hydrogen as its energy source. The Audi-media center had this to say about Audi's attitude towards the future of hydrogen fuel celled vehicles. "Because fuel cell technology-h-tron-presents a whole series of opportunities for sustainable mobility in the medium term, the Audi development engineers are also advancing this technology. The fuel cell is at its best when operated with sustainably produced hydrogen. Audi is developing pioneering solutions with respect to both the technology and production of the fuel." (Audi-Media, 2016) In 2020, Ballard Power Systems announced a new fuel-cell stack developed in connection with Audi. The plan was for the h-tron to be in production by 2025. That was then, this is now. 2022, Audi decided to shelf the hydrogen fuel cell concept. "It's committed to an electric future for passenger cars and four-wheel drives and is effectively putting a line through hydrogen fuel-cell technology. "I don't think the fuel-cell will be relevant for car in general, said Audi head of corporate strategy, Silja Pieh" (Car Expert, 2022)

Volkswagen

"The Volkswagen Group's decision is clear: as a large volume manufacturer, it is focusing on battery-powered electric cars for the masses—even though Volkswagen Group Research is continuing to explore fuel cell technology and Audi, for example, has announced a small-scale hydrogen-powered vehicle for 2021." (Volkswagen Group, 2019) Volkswagen's policy was made clear in 2020 with the following statement.

"Volkswagen has also decided against the technology, with Herbert Dies, the company's chief, telling industry insiders in July: 'It doesn't make a lot of sense at this point to think about bringing hydrogen into passenger cars." (Korn, 2020)

"The current facts prove Volkswagen right. Prof. Maximilian Fichtner, Dep. Director of the Helmholtz Institute Ulm for Electrochemical Energy Storage and designated expert in hydrogen research, lately told the "Wirtschaftswoche", the 'very poor energy efficiency well-to-wheel' of the fuel cell car make sure that battery-powered e-cars...." (Volkswagen Group, 2019)

The Automotive Industry 2035 – Forecasts for the future, studied the expectations of what people want to know in terms of expectations for the next 10 to 15 years. "The study first analyses the reasons for the purchase. Why should customers switch to e-cars? The most likely scenario at the moment is a two-phase model, according to the study: the push-phase and the subsequent pull-phase. In the push phase from today to 2023/2025, manufacturers will push e-mobility. The main reasons for this are the strictCO_2 standards. Added to this are the initially high investment costs. Both of these factors mean that purchase incentives must be set in order to bring e-cars onto the market. In the subsequent pull phase until 2030 and above all until 2035, e-cars will also become more financially interesting for customers." (Volkswagen Group, 2019)

"The most interesting part of the study remains: Which energy has the best efficiency and is the most cost-effective for driving e-cars? Battery of hydrogen operation? With battery-powered e-cars, only eight percent of the energy is lost during transport before the electricity is stored in the batteries of the vehicles. When the electrical energy used to drive the electric motor is converted, another 18 percent is lost. This gives the battery-operated electric car an efficiency level of between70 to 80 percent, depending on the model." (Volkswagen, 2019)

"With the hydrogen-powered electric car, the losses are significantly greater: 45 percent of the energy is already lost during the production of hydrogen through electrolysis. Of this remaining 55 percent of the original energy, another 55 percent is lost when hydrogen is converted into electricity in the vehicle. This means that the hydrogen-powered electric car only achieves an efficiency of between 25 to 35 percent, depending on the model. For the sake of completeness: when alternatives fuels are burned, the efficiency is even worse: only 10 to 20 percent overall efficiency." (Volkswagen, 2019)

"The conclusion is clear: fuel cell e-cars have many advantages (range, fast refueling, no heavy battery on board), but one decisive disadvantage: It is comparatively inefficient both in terms of efficiency and cost. 'No sustainable economy can afford to use twice as much renewable energy to drive fuel cell cars instead of battery-powered vehicles,' says Dietmar Voggnereiter, head of the study. Hydrogen could only be used in niches, in trucks and buses, and over long distances. Battery weight, range and fueling time play a decisive role here. It increases extremely with increasing capacity, which makes batteries uninteresting even for trucks. In addition, existing truck filling stations could be converted to a hydrogen filling station network with manageable effort due to their lower numbers." (Volkswagen, 2019)

That study and definitive decision was in 2019. In contrast, in 2021, Volkswagen applied for a patent along with the German company, Kraftwerk Group, on a new fuel cell vehicle. It was announced that Volkswagen was developing a hydrogen car with a 1,250-mile range. "Volkswagen is developing a new fuel cell hydrogen car that is intended to be substantially cheaper than the currently available H2 vehicles. They are aiming to roll out a model that will have a range of 2,000 kilometers per fuel tank." (Moore, 2022) This seems to be a reversal of Volkswagen's stance two years before. The big advantage was the development of a ceramic membrane in place of the plastic (polymer) membrane in the current hydrogen fuel cell. "Key among the technologies being pursued by VW is a ceramic membrane. Described as being cheaper than the polymer membrane used by Hyundai and Toyota, it is considered crucial to the future volume production of hydrogen fuel cells for automotive applications." (Kable, 2022) The advantage is that the ceramic membrane is cheaper to produce and does not require any type of platinum.

BMW

"In 04 the BMW H2R set nine world speed records for internal combustion powered vehicles using hydrogen as a fuel. The H2R has a 0–60 acceleration in under 6 seconds and a top speed of 187.62 mph (301.95 km/h) from 232 horsepower (173 kW)" (Smartcharger, 2020)

In 2013, BMW and Toyota formed an alliance to co-develop a drive system using hydrogen fuel cell technology. In 2015, BMW began testing a small fleet of BMW 5 Series vehicles using hydrogen fuel cells. In 2016, BMW and Toyota signed an agreement to work together on future generations of hydrogen fuel cell vehicles. BMW Group and Toyota teamed up in 2017 along with eleven leading companies in energy to form the Hydrogen Council. By 2019 the Hydrogen Council had grown to 60 companies. "In the future, the hydrogen fuel cell drive can be an attractive alternative to battery-electric vehicles. In particular for customers who do not have their own access to electric charging infrastructure and who often drive long distances, BMW said, with a sufficient refueling infrastructure, hydrogen vehicles offer great flexibility, as the full range is available again after a short refueling process of around four minutes regardless of temperature conditions." (Greencar, 2020) In 2020, BMW reaffirmed its commitment to hydrogen fuel cell technBMW estimates that by 2030 fuel cell vehicles will be on a par with BEVs. ology. "When the market is ready, the hydrogen fuel cell technology could become the fourth pillar of the strategic approach in the long-term. BMW Group is confident that convental engines will continue to co-exist with electric, plug-in hybrid and potentially, hydrogen fuel cell alternatives for a long time." (Dorofte, 2020) "The road to free-emission mobility is taken on systematically and gradually, by carefully considering the differing market and client requirements as part of the company's already established 'Power of Choice' strategy."(Dorofte, 2020) "We are convinced that various alternative powertrain systems will exist alongside one another in future, as there is no single solution that addresses the full spectrum of customers' mobility requirement worldwide," Klaus Frohlich, member of the board of management of BMW AG, said last fall. "The hydrogen fuel cell technology could quite feasibly become the fourth pillar of our powertrain portfolio in the long term. The upper-end models in our extremely popular X family, particularly suitable candidates here." (Ramey, 2020)

"In 2022, the BMW Group is planning to present the next generation of hydrogen fuel cell electric drive systems in a small-series vehicle based on the current BMW X5. The BMWi Hydrogen NEXT provides an initial glimpse of what this model has in store. The BMW Group would start offering fuel cell vehicles for customers in 2025 at the earliest, but the timing very much depends on market requirements and overall conditions." (BMWGroup, 2019) "Goodbye electric car: BMW wants to produce hydrogen cars in series." (Automotive, 2022) "BMW is hydrogen's biggest proponent among Germany's carmakers, charting a path to a mass-market model around 2030. (Carey, 2021)

"When it comes to eco-friendly vehicles, most people think batteries over hydrogen. A new partnership between BMW and Toyota, however, may be giving the latter approach a desperately needed publicity boost." (Chen, 2022) "Greater use of fuel cells for BMW vehicles could help reduce its need for raw materials like lithium and cobalt. Its hydrogen fuel-cell system uses aluminum, steel and platinum, all of which are recyclable." (Shepford, 2022) BMW is teaming up with Toyota to produce hydrogen fuel cell electric vehicles. BMW sales chief, Pieter Nota had this to say regarding partnering with Toyota. "A BMW official told Nikkei Asia during an interview that the Germans are seeing obvious benefits of hydrogen fuel cell vehicles, especially in the case of larger SUVs. And Toyota could be the most-obvious choice for a partner because the Japanese have been developing the technology since the early 1990s." (HT Auto Desk, 2022)

"This partnership between the two car makers means they're both committed to pursuing hydrogen fuel cells as an alternative to the more popular battery-powered electric vehicles." (Chen, 2022)

The BMW SUV will be called the iX5. The iX5 will have a single electric motor that will send power to the rear axle. It will not be an all-wheel drive vehicle. "Since the 2000s, BMW has blazed a trail in hydrogen fuel-cell technology despite limitations. The iX5 hydrogen stands as a testament to BMW's vision. This hydrogen-powered SUV boasts a 313-mile range, effectively addressing rage anxiety. Moreover, its rapid refueling time of 3–4 minutes rivals that of gasoline vehicles, offering a stark contrast to the lengthy charging times of battery-electric counterparts." (Tariq, 2024) "BMW is using the iX5 Hydrogen as a real-world laboratory, collecting crucial data on fuel cell vehicle (FCV) usage and driving the growth of hydrogen infrastructure. The pilot program aims to spur investment in hydrogen refueling stations and production facilities, paving the way for wider adoption of FCVs." (Tariq, 2024)

Stellantis

"Stellantis was the name given to the merger of Fiat-Chrysler (FCA) and Peugeot parent company Groupe PSA, which finalized in January 2021. It encompasses more than a dozen car brands and related companies." (Zachariah, 2021) Stellantis is huge and is the parent company of: Abarth, Alfa Romeo, Chrysler, Citroen, Dodge, DS Automobiles, Fiat, Jeep, Lancia, Maserati, Opel, Peugeot, Ram, and Vauxhall car manufacturers. Stellantis has also acquired a third interest in Symbio, a leader inzero-emission hydrogen mobility. In addition to Stellanti, Michelin and Faurecia each hold a third interest in Symbio. "Stellantis has developed a Hydrogen Fuel Cell Zero Emission solution which combines the advantages of hydrogen fuel cells and electric battery technology n a Fuel Cell Electric Vehicle (FCEV). The solution is particularly suited to the needs of light commercial vehicle (LCV)customers requiring long-range, fast refueling, and zero emissions…all without compromising payload capacity." (Stellantis, 2021) In 2021, Stellantis produced hydrogen fuel cell mid-size transport vehicles called Citreon Jumpy, Peugeot Expert and Opel Vivare featuring hydrogen fuel cell technology. The mid-size Citroen Jumpy, Peugeot Expert, and Opel Vivaro vans were chosen as the launch models for this technology, enabling rapid adaptation of the technology itself and rapid integration with our existing production processes." (Stellantis, 2021) Stellantis worked in partnership with Faurecia and Symbio, to develop hydrogen storage and fuel stack technology. Stellantis has developed three versions of HFC power, the range-extender, Stellantis mid-power concept and the full power version. The Stellantis mid-power concept has three advantages, enables smart packaging, no compromise in terms of performance, and the battery covers power requirements for acceleration in terms of performance.

Alfa Romeo

Stellantis intends for Alfa Romeo will be all-electric by the year 2027. At present, as of this writing in June 2023, Alfa Romeo has no electric vehicles including no hydrogen fuel cell cars. "According to Autoblog, Alfa Romeo's first EV might be slated to be a crossover or SUV, which means it could share a platform with the Jeep EVs." (Motor Biscuit, 2021) Automotive designer Daniel Kemntz, Alfa Romeo revealed a hydrogen-powered Alfa Romeo P7 hypercar concept in 2023. "Why hydrogen? Kemnitz believes that new vehicles must have either a hydrogen or electric power source, with the former bridging the link between the emotional past as well as the visionary future. Hydrogen fuel cells are essentially made through the REDOX (reduction/oxidation) reaction between hydrogen and oxygen. This process converts the energy into electricity, using hydrogen as the fuel and then combining it with oxygen to produce an electrochemical cell." (Smith, 2023)

Dodge

"Recently, Stellantis made a big announcement regarding Dodge. They are producing an EV that is strictly about performance. Dodge's muscle car heritage should be front and center in the new EV that promises performance over most anything else and should rival the Tesla Model S Plaid as the new standard in EV performance." (Motor Biscuit, 2021)

Ram trucks have been working on going electric with their trucks. Ram has stated that they anticipate their entire lineup will be electrified by 2030. Question? Are Ram trucks considering working on hydrogen fuel cell trucks for their heavy-duty trucks? Hydrogen fuel cell vehicles are said to be better suited for heavy-duty loads. In 2022, Stellantis, parent company for Ram trucks announce this. "The first hydrogen-powered Ram Heavy Duty pickup truck was announced during Stellantis' Dare Forward 2030 presentation earlier this week. The revelation was made alongside the first Jeep fully electric SUV due to launch in early 2023 and the new Ram 1500 BEV pickup truck arriving in 2024." (Accardi, 2022) The expected name for this new Ram truck is the "Revolution." "Stellantis views fuel cell electric vehicles as 'well suited to the needs of commercial vehicle customers requiring long-range, fast refueling and zero-emission without compromising payload capacity.' It's likely the company will expand hydrogen fuel-cell technology to the hydrogen-powered Ram HD instead of pursuing an internal combustion engine on liquefied hydrogen." (Accardi, 2022)

Jeep

"The new Wrangler EV features all the off-road features people love about it while making it both environmentally friendly and wallet-friendly. Jeep is also looking into hydrogen fuel cell technology and promises that the EVs will be even more capable off-road than gasoline-powered Jeeps." (Motor Biscuit, 2021) The design center at Chrysler Advanced Product Design has developed the Jeep Treo. The Treo was designed to appeal to the younger buyer. Power is provided by two electric motors, one mounted on each axle, to give four-wheel drive driven by a hydrogen fuel cell mounted under the floor. The Jeep Treo would be a zero-emission vehicle.

Chrysler

As early as 2001, Chrysler was working on a hydrogen fuel cell vehicle. In 2001, Chrysler introduced the Natrium, a fuel cell version of the Chrysler Town and Country minivan. The Natrium was shown at the 2001 Electric Vehicle Association of the Americas in Sacramento, California. Chrysler said they had been working on the Natrium for 18 months. The hydrogen is derived from sodium borohydride, a chemical that is similar to the laundry detergent, borax. One drawback was the need for filling stations to have two tanks, one for with fresh fuel and the other to dump the spent fuel. The Natrium never reached production. Sodium borohydride didn't pan out as a power source for hydrogen fuel cell vehicles.

Harold Wester, the Chief Technology Officer for Fiat-Chrysler said, "Hydrogen fuel cells are a better choice for future power trains than batteries." (Grrencars, 2020) Wester is also the CEO of Maserati and Alfa Romeo, said he was discussing possible alternative for these two vehicles as well. "Wester also claimed that continued use of fossil fuels to generate electricity ultimately makes battery-electric cars less green than fuel-cell vehicles." (Greencars, 2020)

Fiat

Fiat has a history of hydrogen fuel cell concepts going back to 2004. Working with Nuvera Fuel Cells Inc., Fiat called this early concept the Panda. The hydrogen stack and fuel tank was stored under the floor pan. In 2006 a new Panda concept was introduced. "The 2006 Fiat Panda Hydrogen Car uses a full power system, which means it lacks a drive battery for the accumulation of electrical energy. The Panda Concept receives enough energy directly from hydrogen tank to fuel cell to deliver the needed electricity to its high-torque electrical motors." (Hydrogen Cars Now, 2023)

In 2010, the Panda concept was upgraded again.

Peugeot

"Pushing ever further its requirement to drastically reduce the local emissions of its vehicles, PEUGEOT is becoming one of the very first manufacturers to offer series production, from 2021, in the compact utility van segment an electric version powered by a hydrogen fuel cell, consisting of:

A fuel cell producing the electricity needed to propel the vehicle thanks to the hydrogen on board the tank.

A rechargeable lithium-ion high voltage battery, with a capacity of 10.5 kWh rechargeable from the electrical network, which also powers the electric motor in certain driving phases." (Peugeot, 2023)

"Peugeot has become one of the first manufacturers to offer in series production, from 2021 onwards in the compact utility van segment, an electric version powered by a hydrogen fuel cell in addition to its battery-electric version." (Greencar, 2021) This is historic. The Peugeot e-EXPERT had a hydrogen fuel cell in the engine compartment that produced electricity when hydrogen was mixed with oxygen, with water vapor as exhaust. The fuel cell supplied the energy directly to the electric motor. It also had a permanent electric magnet motor as a drive chain. As a back up there was a high-voltage lithium-ion battery located under the front seats. The e-EXPERFT in addition had 3 hydrogen storage tanks located under the floor. The e-EXPERT could be fueled in three minutes. The e-EXPERFT had three driving modes, eco, normal and power. In addition, there were two braking modes.

"Linda Jackson, CEO of the Peugeot brand, said: With the e-EXPERT hydrogen, Peugeot is taking the lead in a future zero-emission technology that is particularly relevant to the light commercial vehicle market: hydrogen electric technology allows intensive daily use without the need for recharging, a decisive advantage for professionals who, in a single day, have to cover several hundred kilometres on the motorway and then enter emission-restricted urban areas. Hydrogen technology developed by Stellantis and implemented in our Peugeot e-EXPERT Hydrogen makes such uses possible." (Greencar, 2021)

"The electric vehicle market has exploded due to concerns about climate change, but there are concerns about the environmental impact of electric drivetrains. As an alternative, hydrogen-powered vehicles are emerging, with benefits like longer ranges and emissions that are nothing more than water or vapor. However, the cost of hydrogen-powered drivetrains is currently high, and technological advances are needed to make it more accessible." (Arsenale, 2023) Looking to the future also leads to concept cars. Zhe Huang, a design student has proposed a "no concept," Peugeot. "The car has a minimalist appearance and a intriguing visual balance, making it stand out from other concept cars that are often impractical or too futuristic." (Arsenale, 2023) "The Cyberpunk-inspired design of the Peugeot 'No Concept' flows smoothly from the front to the rear, with a concave side profile that adds a muscular appeal. The car's aerodynamic design is optimized to reduce drag coefficient for better fuel efficiency and overall range." (Arsenale, 2023)

Peugeot has also produced a very practical FCEV in the form of a fuel cell powered fire engine. "The H2O is an electric vehicle with batteries, fitted with an auxiliary fuel cell which provides a continuous source of electrical energy to supply various emergency items of equipment such as the pumps, smoke extractors, communications systems, and electric sockets." (Hydrogen House Project, 2022) The fire truck also equipped with proximity sensors, radar system, extension ladders, a water tank, and room for two in the front seat. The interior is loaded with a touch screen, GPS system, telephone and a second screen that projects maps.

Opel

In 2006 GM in cooperation with Opel produced the HydroGen3 a small compact hydrogen fuel cell vehicle that was approved for testing in Japan. The hydrogen3 was a 5 passenger, front wheel drive protype that had an estimated 250-mile range. The first production version of the Vivaro was introduced in 2021 and has already been delivered.

Opel has embraced hydrogen fuel cell technology in its small Vivaro commercial utility van. The Vivaro was to be launched in Europe in 2023. The Vivaro would have the same cargo space as the diesel and electric models and have an estimated 249-mile range. The van will appear on the streets of Germany. "Hydrogen could be the central element of an integrated and efficient energy system of the future, free of fossil fuels, said Michael Lohscheller, Opel's CEO. We have more than 20 years of experience in hydrogen fuel cell vehicle technology." (Gnaticov, 2021) "The clever concept combines the advantages of hydrogen fuel cell propulsion with the versatility and capabilities of our best-selling light commercial vehicle, said Opel CEO Hochgeschurtz," (Stellantis, 2021)

Maserati

Maserati's fuel cell history goes back to 2015. At that time Maserati was under the parent group Chrysler-Fiat. Harald Wester, CEO of Fiat-Chrysler at the time may have been predicting the future in this interview with Motor

Trend. "You need to bring the fuel cells not only to a certain technical and performance level so that you can keep the buffer battery relatively low. You also need to bring it to an affordable economic level. But then it's the future." Wester told Motor Trend." (Stevens, 2015)

Renault

Renault introduced a new concept car called Scenic Vision that has a purported range of 497 miles plus many unique features especially in terms of economy and the environment. How are plastic milk bottles involved in the construction of this concept? The Scenic vision features eco-friendly construction. The car is made of 95% recyclable materials. Some body parts and the fuel tank are made from 100% recycled material from aviation industry scrap. 30% of the car's plastic components are bio-sourced. The floor is made from 100% recycled plastic from milk bottles and plastic pipes. The Scenic Vision uses no rare-earth elements.

The Renault Scenic Vision is loaded with other distinctive features. One, is the risk assessment which can detect risky driving behaviors and encourage defensive driving practices. There is also a camera system installed in the front end that enhances the driver's visibility up to 24%. The interior has a large display screen and seven smaller screens. Plus, the steering wheel looks like a Formula 1 car with a screen in its center. The plan was to market it by 2024.

But what about the power unit? The Scenic Vision is powered by an electric-hydrogen hybrid engine. The fuel tank is located to the front and takes about five minutes to fill. The vehicle can run on batteries that are twice as light as regular electric car batteries, saving weight. "The H2 Tech technology is based on the 'range extender' technology, which makes it possible to carry a battery that is twice as light, for the same range, thus contributing to decarbonization beyond the electrification of the vehicle.' The manufacturer said." (Matalucci, 2022) "The broad idea is that the Scenic Vision's hydrogen fuel cell would extend the vehicle's range during longer trips. 'In 2030 and beyond, once the network of hydrogen stations is large enough, you will be able to drive up to 800 km (a little over 497 miles) without stopping to charge the battery," Renault3 said." (Franhgoul, 2022)

Toyota

Toyota has been working on fuel cell vehicles since 1992. "Toyota was one company that truly believed in the hydrogen future and produced the very credible FCV-R concept in 2011 that developed into the Mirai, which became commercially available in 2015. In 2017 Toyota's Mirai was introduced in dealerships in Southern California. A second generation will be released in 2021." (Morris, 2020) "Toyota may have been one of the first major car makers to test the waters with hydrogen-powered EVs, but it's clear that the battery-powered EVs are dominating the market." Chen, 2022) In contrast, Toyota's first all-electric EV, the bZ4X EV, has turned out to be a total failure. Toyota has even offered a buy-back program. Only 2,700 units were sold globally. "Considering Toyota's reputation for reliability and durability, it's surprising that it bungled the launch of its first accessible EV so badly. Toyota's failure is only compounded by the fact that it's falling behind in the EV sector, like GM, Hyundai, and BMW." (Chen, 2022) The recall has also affected the Subaru Solterra which shares the platform with the bZ4X.

"Toyota sees tremendous upside in fuel cell technology, which it has been perfecting for 25 years. More than 6,500 Mirais have been sold or leased in California since its launch in 2015. The second generation Mirai, on sale next month in San Francisco and Los Angeles, can store more hydrogen than its predecessor, giving the sleek sedan a 30% increase in range." (Korn, 2020)

In December 2022, Toyota announced plans to develop a hydrogen fuel celled pickup truck called the Hilux. "Toyota and Hino are currently developing a series of hydrogen fuel cell electric trucks." (WHA, 2023) Toyota is developing the Hilux Hydrogen truck with funding from the UK government. Limited production was scheduled for 2023 but has not been met.

"Toyota's hydrogen Hilux prototype remains a fascinating glimpse into the future of zero-emission pickups, but concrete information remains scarce. The prototype aimed to test the feasibility of adapting second-generation fuel cell technology from the Mirai sedan into a rugged truck platform." (Tariq, 2024) "Along with the Mirai and the new Hilux, Toyota has also been working on hydrogen buses and heavy-duty trucks. If larger transport vehicles transition to hydrogen alongside commuter cars, we might reach our zero-emissions goals even faster." (Fast Tech, 2023) Toyota has also worked on a hydrogen fuel cell bus called the Sora along with heavy duty truck prototypes.

Toyota has also worked on a hydrogen powered rally car. The Toyota GR Yaris H2 is setting the stage for hydrogen combustion in the competitive rally sport. "Toyota took the automotive world by storm in December 2021 with the unveiling of the GR Yaris H2. This innovative rally car marked a significant step forward in hydrogen combustion technology, offering a unique alternative to fuel-cell electric vehicles. Despite the excitement, Toyota remains tight-lipped about specific performance figure like horsepower and range, leaving the world in anticipation." (Tariq, 2024) The Yaris is a rally proven platform, renowned for its performance and agility. The GR Yaris H2 uses a hydrogen combustion engine not fuel-cell power. The hydrogen version has the potential for increased horsepower. "It features a modified turbo-charged three-cylinder engine from the standard GR Yaris and is expected to generate around 268 horsepower." (Fast Tech, 2023) The car was tested in the Ypres Rally in 2022.

Honda

Honda fuel cell technology milestones include:

- Honda FCX was the first EPA- and CARB-certified fuel cell vehicle (July 2002)

- FCX was the first world's first production fuel-cell vehicle, introduced to the U.S. and Japan (December 2002)

- First fuel-cell vehicle to start and operate in sub-freezing temperatures (2003)

- First fuel-cell vehicle leased to an individual customer (July 2025)

- First manufacturer to build and produce a dedicated fuel-cell vehicle (FCX Clarity) on a production line specifically made for fuel-cell vehicles (2008)

- First automaker to create a fuel-cell vehicle dealer network (2008)

- Honda FCX Clarity Plug-in Hybrid served as the pace car fotr the 2008 Indy Japan 300.

- 2011 the Honda FCX Clarity also served as the pace-car at the St. Petersburg Honda Grand Prix.

- Honda entered into a long-term collaborative agreement with General Motors to co-develop the next generation of fuel-cell systems and hydrogen storage technologies (June 2013)

- As the next progression of Honda FCEVs, Honda introduced the Honda Clarity Fuel cell (April 2017)

- In 2020 Honda announced changes in the Honda Clarity Fuel Cell, would allow it to run better in colder temperatures.

- Honda and GM announced a joint venture (Fuel Cell Systems Manufacturing) working toward mass production of lower cost, compact and high-performance fuel-cell systems (2017) (Honda News, 2023)

Honda was among the pioneers in fuel cells, showing a prototype car in 1998, and its first production car the Clarity in 2002." At its debut, the

Honda FCX Clarity was the first hydrogen fuel cell vehicle offered to the public." (Uwaoma, 2022) In 2017 the Clarity became a production hydrogen fuel-cell car that was available for purchase or lease in Southern California. Because of the lack of sales and the lack of infrastructure for refueling the Clarity has been discontinued as of 2022. "The Honda Clarity was a great car, but the absence of reliable hydrogen infrastructure would certainly be a contributing factor to its demise." (Uwaoma, 2022)

"Despite the failure of the Clarity model, the Japanese auto giant is still pushing for fuel-cell automobiles." (Martin, 2024) "This is despite a disappointing start for Honda's foray into hydrogen-powered cars, the Clarity Fuel Cell, which inly saw about 1,900 units sold worldwide, leading to its discontinuation in 2021." (Martin, 2024) Honda's thinking is that that fuel-cell cars will need more time but will take-off by 2040. "What I have in my mind is that the (battery) EV era-comes first, and the next phase is fuel-cell cars., said Katsushi Inoue, CEO of Honda India as of the beginning of this year and former Europe chief, in his capacity leading the company's electrification efforts." (Martin, 2024)

"Since 2013, Honda has been working with GM on the joint development of the next-generation fuel cell system' Honda explained, confirming that it plans to launch a new fuel cell electric car in 2024. 'Equipped with the next-generation fuel cell system jointly developed with GM." (Casey, 2023) In February of 2023, the Associated Press announced: "Honda's next hydrogen fuel cell vehicle to get stack developed with GM." "GM and Honda have begun commercial production of hydrogen fuel cell systems in a step toward offering alternative zero-emissions solutions beyond battery-electric vehicles." (Wayland, 2024)

Estimates are that the new fuel cell system would cost just a third as much as the fuel cell stack in the Clarity. Honda expects to have a new fuel cell powered vehicle for 2024, that will have a new fuel-cell technology developed in cooperation with General Motors. The goal is to sell 2000 units of the next generation fuel cell system by 2025. This new fuel cell vehicle will be a SUV. Honda is also looking to expand the use of hydrogen in trucks and construction equipment. "Honda said it sees fuel cell systems as vital players in a carbon-neutral society and it doesn't just want to put them in cars. The automaker is searching for partners to test them in commercial vehicles, stationary power stations and construction machinery." (Blanco, 2023) Honda considers fuel cells as more efficient in producing energy than electric batteries. "Since 2008, Honda has diligently pursued hydrogen technology, and the 2025 CR-V Hydrogen represents a potential turning point in their strategy. This innovative iteration takes a significant step away from its predecessor, the Clarity Fuel Cell, by integrating a battery pack alongside the fuel cell system. This unlocks bidirectional charging capabilities, allowing the CR-V Hydrogen to function not only as a vehicle but also as a versatile backup power source, significantly expanding its potential uses." (Tariq, 2024)

Kia

Kia has been slow to develop a hydrogen fuel-cell vehicle. Present plans are looking at 2028 for the introduction. Kia does have an agreement with Hyundai for parts. The Vision FK, which is a Hyundai sports car concept is based on the Kia Stinger. "During the 'Hydrogen Mobility Show' in South Korea, Song Ho-Sung, president, and CEO of Kia Motors, told reporters that Kia will focus on using hydrogen t4echnology for the military, before using it in passenger vehicles starting from 2028." (Pappas, 2021) The expectation is to convert their entire military fleet to hydrogen fuel cell vehicles. Part of the reasoning is that it would take less time to refuel a military vehicle with hydrogen than to recharge a lithium-ion battery vehicle. Kia is also planning to use hydrogen fuel-cell power as an energy source at all their worksites.

Kia has introduced a concept SUV called EV9. "To demonstrate Kia's emphasis on more upcycling, the EV9's seats and flooring were made of recycled fishing nets, while seats fabric were made of recycled plastic bottles, and flooring is made of wood." (Halvorson, 2021) The interior will also have vegan leather. 0Kia plans to phase out the use of leather in its vehicles.

Mazda

Mazda's history with hydrogen fuel cell technology goes all the way back to 1991. Mazda fabricated its first fuel cell vehicle powered by a PEFC (Polymer Electrolyte Fuel Cell) supplied by Ballard Power Systems. In 2001, Mazda introduced the Premacy FC-EV powered by a methanol reformer fuel cell system. The Japanese Minister of Land Infrastructure gave permission to Mazda to test drive the Premacy on the Japanese highways. "Mazda has been working with Ford Motor Company on research and development of the FC-EV. In April 1998, through its close relationship with Ford, Mazda began participating in the fuel cell alliance established by Ford Motor Company, Daimler Chrysler and Ballard Power Systems. The Premacy FC-EV has been developed in cooperation with Ford's THINK group." (Mazda, 2001) THINK was Ford's research and development division. In 2010, Mazda introduced the Premacy Hydrogen RE Hybrid to be leased by Mazda. The hybrid's power was delivered to the wheels by electric motors.

The Premacy Hydrogen Hybrid could run on gasoline as well as hydrogen.

In 2021, there were rumors that Mazda was developing a hydrogen powered rotary engine. Note, that's hydrogen not hydrogen fuel cell power. Mazda stopped building the Mazda 8 rotary engine back in 2012. Mazda is developing the MX-30 electric crossover. "One of hydrogen's weak points is that it tends to ignite at heat spots inside the cylinders. Best Car noted that there are no heat spots in a Wankel engine, because it uses rotors rather pistons, so it's well suited to burning hydrogen." (Glon, 2021) Mazda confirmed rumors that the rotary engine MX-30 would return in late 2022. However, Mazda has back tracked on that introduction date.

"The hydrogen rotary engine is extremely eco-friendly and perfect for a society in which people can continue to enjoy Zoom-Zoom driving while caring for the Earth." (Environmental technology, 2022) "The RENESIS hydrogen rotary engine employs direct injection, with an electronically-controlled hydrogen gas injector. This system draws in air from a side port and injects hydrogen directly into the intake chamber with an electronically-controlled hydrogen gas injector installed on the top of the rotor housing.". (Environmental Technology, 2022)

Nissan

Hydrogen fuel cell power, technology and vehicles are not new to Nissan. Dating back to 2003 Nissan produced a compact SUV, the X-TRAIL FCV. In 2004, Nissan introduced an updated model that featured greater acceleration, higher hydrogen storage capacity, a longer cruising range, a modified, smaller more compact fuel stack and personal conveniences, like passenger comfort and increased storage space. "The fuel stack is the heart of the FCEV, where hydrogen and oxygen found in the atmosphere is caused to electrochemically react in order to generate electricity. Further size reduction, high output, and cost reduction is possible if the power density of the stack is improved." (NissanMotors, 2007)

"The new technology, dubbed an e-bio fuel cell, aims to combat a common hurdle to deploying traditional hydrogen fuel cell vehicles; the lack of a hydrogen fueling infrastructure. Nissan's system uses bio-ethanol, derived from renewable crops such as corn or sugarcane, and that refueling infrastructure already largely exists." (Grimel, 2016)

In 2016 Nissan was working on replacing hydrogen with ethanol as a fuel source. "Nissan has developed a new kind of fuel cell drivetrain for cars that taps an onboard tank of ethanol instead of pressurized hydrogen, delivering a cheaper and safer ride that it says is more user friendly." (Grimel, 2016) "The technology that Nissan plans to use in its future fuel-cell vehicles uses heat to reform ethanol into hydrogen to feed what is known as a solid oxide fuel cell, or SOFC." (Voelcker, 2016) This SOFC engine would use oxygen ions in place of protons to move through the electrolyte to produce electricity with water as a byproduct. "Hydrogen produced in the reformer is fed into the solid-state fuel cell, which generates electricity at a relatively steady rate to supply power to the electric motor driving the wheels, through a battery that handles peak power demands and stores regenerated energy. Heat from the fuel cell is used in the reformer, Nissan said, creating a highly efficient system." (Voelker, 2016)

Hyundai

"Hyundai is one of the leading companies in the field of fuel cell technology and has been producing hydrogen-powered cars since 2013." (Carscoops, 2020) In 2020, Hyundai introduced a long-term strategy with a fuel cell brand called HTWO. "With HTWO, Hyundai Motor Group is stepping up efforts for the development of a next-generation hydrogen fuel cell system that can be applied to various forms of mobility such as UAM, automobiles, vessels and trains, 'Hyundai said in an announcement late Thursday." Hurd, 2020) "Not only will the next-generation fuel cell system be available for many different mobility products and services, it will deliver enhanced performance and durability at an affordable price in a lighter architecture with enhanced energy density. With its next-generation fuel cell system, the group aims to offer a highly efficient and diversified lineup of hydrogen-powered vehicles." (Hurd, 2020) "Hyundai has also made significant progress in hydrogen fuel-cell space. It was the first automaker to bring out a production crossover when it unleashed the Tucson Fuel Cell in select dealerships in Southern California." (Pleskot, 2015) "The Hyundai Nexo represents a small step for the South Korean automaker, and a giant leap for public acceptance of hydrogen fuel cell vehicles. It's the first fuel cell vehicle to go through the full battery of Insurance Institute for Highway Safety (IIHS) crash tests—and it did well." (Glon, 2020)

In 2020 Hyundai introduced the first production hydrogen fuel cell SUV the Nexo. In 2020, there were only three hydrogen fuel cell cars, the Toyota Mirai, Honda Clarity and the SUV, Nexo from Hyundai. "Without any hyperbole, the Nexo is one of the quietest and comfiest cars this side of a Mercedes-Benz S-Class. The suspension damping is sublime, even with the 19-inch wheels, and it's not unsettled by even the biggest potholes and bumps. A little bit of wind noise and tire noise makes its way into the Cain, but otherwise the Nexo is dead silent. It's an extremely calming car both to drive and to be driven in, and even after long journeys I get out feeling refreshed." (Golson, 2020) In 2024, Hyundai had planned to introduce a third generation fuel stack and an impressive range of nearly 500 miles. "One of the most notable aspects of the Nexo is its hydrogen fuel cell powertrain. This clean and efficient technology offers an impressive range of 354 to 380 miles on a single fill. Beyond its eco-friendly credentials, the Nero impresses with its technology and comfort features." (Tariq, 2024)

The N Vision 74, is a hydrogen fuel cell muscle car from Hyundai "The N Vision 74 was touted by Hyundai as a rolling lab, testing various technologies in the hopes that they could reach production, whether that performance fuel-cell powertrain will see production remains to be seen." (Biermann, 2023) As of 2022, there is no plan to put this pony car into production. There are those that believe that hydrogen fuel cells will supplement EVs in the future. The N Vision 74's core structure is actually built on the Kia Stinger. "The N 74's tech suggests there will be some serious high-performance models in Hyundai's future." (Duff, 2022) "The Ioniq 5 is an electric vehicle with an N performance version on the way. The N Vision 74 went a different route, eschewing pure electrification for a hydrogen fuel-cell system combined with twin electric motors and a 62.4-kWh battery. This system generates a combined 679 horsepower and a 663 lb-ft of torque, all sent to the rear wheels exclusively (the original Pony was also RWD)" (Biermann, 2023)

"In 2020, Hyundai released the XCIENT Fuel Cell—an achievement it calls 'the world's first mass-produced heavy duty fuel cell truck line.' The XCIENT is a zero-emissions cargo vehicle with more than 400km of driving range. Already on the road in Switzerland, it'll likely transform the overland shipping industry as it reaches more locations." (Fast Tech, 2023)

Ineos

Introducing a new automobile manufacturer, Ineos. Ineos is a British based company that is a major producer of hydrogen. Hydrogen fuel cell car development would be a natural extension of the company and open a huge new market. Ineos will be based in the United Kingdom and will produce a new vehicle called the Grenadier. "Over a friendly pint at the Grenadier pub in London, car enthusiast and experienced adventurer Jim Radcliffe, INEOS Chairman, identified a gap in the market for a stripped-back, utilitarian 4 x 4. This became our vision—to build a no-nonsense 4 x 4, engineered to overcome all conditions and withstand the daily punishment you put it through. For those who depend on a vehicle as a hard-working tool, wherever they are in the world. "Ineos move into the development of a fuel cell electric vehicle and hydrogen ecosystem marks yet another milestone towards sustainable and clean transportation,' said Hyundai fuel exec Saeoon ." (carscoops, 2020)

We're now turning that vision into a reality. More than 200 of the world's best engineers are deep into the task at hand, building the Grenadier from the ground up. With every decision based solely on utility. No design for design's sake. Function over form, every step of the way. Built on purpose." (Ineos, 2020) You may have noticed the SUV has been named for the Grenadier pub. The Grenadier was revealed to the public for the first time in July 2020. The Grenadier will be sold in two versions, a four door SUV and a four-door pick-up truck. The Grenadier resembles the Land Rover Defender with more up dated design with rounded corners. "When in 2015 Jaquar Land Rover announced that after 67 years it was ending production of the iconic Land Rover Defender, British billionaire Jim Radcliffe saw an opportunity. Radcliffe knew JLR was working on a high-tech successor to the Defender, and he figured it would be a much more upmarket vehicle than the old one, leaving an obvious hole in the market. To fill it, Radcliffe tried to buy the old Defender tooling, but JLR refused to sell. Undaunted, he decided to create his own Defender-style off-roader instead." (MacKenzie, 2020) The

Grenadier will be a little wider and shorter than the Defender. Initially the Grenadier will be powered by a conventional gas-powered BMW engine.

However, plans are for the Grenadier to be offered in a fuel cell version using Hyundai hydrogen fuel cell power. In November 2020 it was announced that Ineos would work with Hyundai on a long-term plan to develop a reliable source of hydrogen fuel in Europe and to use Hyundai fuel cell technology for their vehicles. "The world's first SUV fueled by "clean" hydrogen power should be on its way to customers within the next few years. This is a step on from the familiar battery propelled electric car, as it will not need to be plugged in and can generate its electricity on the move." (Chapman, 2020) "Peter Williams, technology director of Ineos, said: 'The agreement between Ineos and Hyundai presents both companies with new opportunities to extend a leading role in the clean hydrogen economy." (Chapman, 2020) The Grenadier production has been pushed back until 2027 due to the lack of infrastructure.

Mercedes

Hydrogen fuel cell technology is not new to Mercedes. Mercedes has been working with hydrogen fuel cell development for over thirty years. In 2009 Mercedes produced a fuel-cell B class car that was ready for public consumption. This car was built on an existing compact model with just the motor and drive train replaced with the hydrogen fuel cell set up. "This is the first time Mercedes has moved beyond a concept car into a 'proof of concept' car with the tech, and it's integrated the new fuel and drive-train into a B-Class car. The combination of an in-production chassis and the newly polished engine means that the car is actually being produced in sample quantities late in 2009, and a short-run production will result in 200 cars in early 2010, which will be sold to customers in the U.S. And Europe. While that's not mass-production by any means, it's unquestionably a significant step in getting road going fuel-cell cars into public's hands." (Eaton, 2009)

The B-Class Mercedes didn't sacrifice interior space and the famous high quality Mercedes construction. The car was designed to cold start in in wintry weather and was crash tested more than normal to be extra sure about its safety. However, in 2010 there wasn't much in the way of infrastructure for hydrogen fuel cell fueling stations.

Mercedes also worked in 2013 with Ford and Nissan to design a fuel-cell powered SUV. "The idea of the collaboration was to kickstart the production of fuel-cell cars and hydrogen infrastructure. Mercedes-Benz was the only carmaker of the three partners to produce a vehicle in the program." (Berman, 2020) Called the GLC F-Cell prototype. "Mercedes-Benz is setting a further milestone on the road to locally emission-free driving with the handover of the first GLC F-CEL vehicles to selected customers in the German market. The Mercedes-Benz GLC F-CELL is unique world-wide as it features both fuel cells and a battery drive which can be charged externally using plug-in technology. (Mercedes-Benz, The GLC F-Cell with its dual drive featured long electric range, short refueling times and everyday practicality of an SUV made it the perfect vehicle. Regardless of the operating mode the GLC F-Cell featured an energy recovery system which made it possible to recover energy during braking or coasting and to store it in the lithium-ion battery. The GLC F-Cell could be fueled with hydrogen in about 3 minutes from two carbon-fibre fuel tanks built into the vehicle floor. Because of cost the only a few were built. No GLCF-Cell vehicles were ever put up for sale.

In April of 2020 Daimler decided building hydrogen fuel cell passenger cars was just too costly with this announcement. "Daimler's Mercedes-Benz is killing its program to develop passenger cars powered by hydrogen fuel cells. The company has been working on fuel-cell vehicles for more than 30 years—chasing the dream of a zero-emissions car that has a long driving range, three-minute fill-ups, and emits only water vapor. In the end, the company conceded that building hydrogen cars was too costly, about double the expense of an equivalent battery-electric vehicle." (Bermanb, 2020)

A later story denied that the CEO said they were dropping hydrogen fuel cell vehicles.

While it seems Mercedes commitment to hydrogen fuel cell passenger cars is in doubt, Daimler has announced plans to build fuel cell powered transporters. This means that Daimler intends to compete with Nikola and Tesla for the electric drive long haul transporters. Nikola is building both electric battery powered, and fuel cell powered versions of their semi-trucks. Mercedes has taken a different approach with a liquid hydrogen fueled transporter. "Daimler Trucks unveiled an all-new Mercedes-Benz GenH2 Truck hydrogen fuel-cell concept truck, which is envisioned for flexible and demanding long-haul transport." (Kane, 2020) The Daimler truck is expected to have a driving range of over 600 miles on a single fill-up, using liquid hydrogen stored in two stainless steel tanks. The two stainless steel liquid fuel tanks represent a weight saving over the pressurized gaseous hydrogen storage tanks. "The use of liquid hydrogen, instead of gaseous hydrogen, is crucial according to the manufacturer as it increases the overall energy density (smaller and lighter tanks) and allows for more range."(Kane, 2020) "In the effort to catch up to the performance capacities of a diesel truck, the storage of hydrogen is in particular focus for the company: Instead of storing the H2 in gaseous form under high pressure where 700 bar is usually the norm, Daimler is planning to use liquid hydrogen (known as LH2) with its significantly higher energy density." (Randall, 2020) Daimler expects to have a consumer test version ready by 2023. The expectation that production could be possible in 2025. The series production model will have two liquid hydrogen tanks and a powerful fuel cell system that will make its performance comparable to commercial diesel trucks without the pollutants.

Just as other companies have combined efforts like Hyundai and General Motors, Daimler has also entered in an agreement with Volvo. "Daimler Trucks emphasized that it has worked with "one partner" in the development of the fuel cells in April, the Stuttgart-based company and Volvo Trucks signed a declaration if intent to bundle their fuel cell activities in a joint venture. In preparation for this Daimler has already transferred all its own fuel cell activities to the recently founded subsidiary Daimler Truck Fuel Cell GmbH & Co. KG." (Randall, 2020) Also, just as Nikola plans, Daimler trucks will be offered in both an electric battery version and a hydrogen fuel cell version. During the introduction presentation in 2020, Daimler CEO made this comment. "Daimler Truck CEO Martin Daum emphasized that the manufacturer was focusing on both the battery and fuel cell. 'This combination enables us to offer our customers the best vehicle options, depending on the application,' said Daum, 'Battery power will be rather used for lower cargo weights and for shorter distances. Fuel cell power will tend to be the preferred option for heavier loads and longer distances." (Randall, 2020) Also at the introduction the German Minister of Transportation made this comment: "For his part the German Minister of Transport talked about the 'huge potential' he saw in hydrogen. 'That is why we have been funding hydrogen as a transport fuel for over ten years—one current example is the concept truck presented today', said Scheuer. 'will continue to provide strong support to the development of climate-friendly drivetrains and innovations in and for Germany. This will include, but not be limited to, significantly expanding the funding of vehicles." (Randall, 2020)

Ford

In 2005, Ford had a fleet of 30 hydrogen fuel cell powered Ford Focus cars, in a seven-city program to test fuel cell technology. "Ford Motor Company has turned over the keys to five hybrid hydrogen Ford Focus Fuel Cell vehicles in Southeast Michigan this week as part of a five-city 30-car program to conduct real world testing of fuel cell technology. Taylor will get four vehicles and Ann Arbor will receive one. Area residents will notice the Focus Fuel Cell vehicles in their neighborhoods as city employees drive the vehicles for city business or, in Taylor, as the water department travels to read meters. The vehicles will also be part of community events to spread awareness of the demonstration program." (Motor Trend Staff, 2005) The Ford Focus will be powered by a Ballard 902 Fuel Cell.

In August 2007, Ford Motor company made history when their Ford Fusion 999 established a land speed record of 207 mph, the world's first production-based hydrogen fuel cell racecar. "The Ford Fusion Hydrogen 999 is Ford's latest environmental innovation, another step on the road toward commercially viable hydrogen fuel cell vehicles." "The car was designed and built by Ford engineers in collaboration with Ohio State University, Ballard Power Systems and Roush." (Smartcharger, 2020) (The number 999 reflects back to a speed racer developed by Ford in 1912.

In 2017, William Clay Ford, Jr. had this to say: "I believe fuel cells could end the 100-year reign of the internal combustion engine." (Wand, 2017) Besides the hydrogen powered racecar, In 2007 Ford also introduced at the Detroit Auto Show the Ford Airstream Concept car, a hydrogen powered cross-over model in cooperation with camper maker Airstream. The car seems to be forgotten; however, it was a for-runner of the cross-over models that are so popular now in 2022. "The Ford Airstream Concept is not merely a showcase of the automaker's learnings in hydrogen fuel cell technology. It also previewed the imminent popularity of the crossover over humble sedans." (Reyes, 2022)

"Ford finds clear words when it comes to heavy pickups. In other words, vehicles that are intended to pull or carry particularly heavy loads. At Ford, these vehicles fall under the class "Super Duty," which means something like "heavy duty". No electric or hybrid models are planned here. As recently as spring 2022, Ford CEO Jim Farley confirmed this, noting that battery technology simply wasn't ready for heavier trucks at the time, the blog fordauthority.com reports." (Automotive, 2022) The problem is the necessary battery capacity. Huge and heavy battery packs would be needed making the useful weight of these vehicles unprofitable. Ford found that 95 % of their truck owners pull more than 10,000 lb. payloads. "This is a really important segment for our country, and it's more likely to switch to hydrogen fuel cells before it goes all-electric," Farley said at a vehicle presentation in Kentucky. At the moment, according to Ford, only gasoline and diesel really meet the needs of this class of vehicles." (Automotive, 2022) "Heavy pickups poorly suited for electric drive." (Automotive, 2022)

"Hydrogen Ford vehicles will remain niche products for now." (Foote, 2023)

"Hydrogen is funny because it's always been the fuel of tomorrow and has been for the last 25 years,' Ford Jr. said. 'Some of you may recall that we invested in a company called Ballard up in Vancouver many years ago because we were quite bullish on the fuel cell future. But for a variety of reasons, both the fuel cell itself, it's basically an onboard chemical factory, and it was hard to manufacture at scale and in their early days didn't perform terribly well in extreme-temperature think the other issue though, is hydrogen itself. Trying to get true green hydrogen until very recently was almost impossible, particularly to scale it. You could do it in a lab, but often it took more energy to extract the hydrogen than you were actually creating." (Foote, 2023)

Ford announced in 2022 that it will supply 50 F-550 hydrogen fuel cell powered utility trucks to Southern California Gas in a research study on the feasibility of hydrogen fuel cell trucks. "Southern California Gas Co. (SoCalGas) today announced it is working with Ford Motor Company on a demonstration project to reduce commercial fleet emissions by developing a F-550 Super Duty Hydrogen Fuel Cell Electric Truck. This collaboration is part of the U. S. Department of Energy's (DOE) Super Truck 3 program which aims to significantly reduce emissions in medium and heavy-duty trucks. This utility's participation is another step towards its ASPIRE 2045 sustainability goals by working to replace 50% of its over-the-road fleet with clean fuel vehicles by 2025 and operate a 100% zero-emission fleet by 2035." (SoCalGas, 2022) "We are honored to work with Ford on their strategy to help reduce emissions, 'said Neil Navin, vice-president of clean energy innovations at SoCalGas energy innovations at SoCalGas. 'This project is a critical step toward finding real-world solutions to decarbonize heavy duty transportation such as our utility fleet with Ford's H2 Fuel Cell Electric F550." (SoCalGas, 2022) The demonstration project will also include a temporary hydrogen fueling station. "Ford's strategy to reduce carbon emissions across the globe includes investigating multiple technologies that will help us achieve these goals across a broad spectrum of applications. For our wide spectrum of Ford Pro customers, there are application gaps that battery electric vehicles just can't fulfill yet, so we're looking at hydrogen fuel cells to power larger, heavier commercial vehicles while still delivering zero tailpipe emissions." (Green car congress, 2022) With the addition of the 50 F-550 HFC trucks, SoCalGas becomes one of the first utility companies in the nation to start the transitioning to hydrogen powered vehicles.

In May of 2023, Ford announced that they will begin a three-year trial for Ford e-transit trucks powered by hydrogen fuel-cells in the United Kingdom. Ford believes the e-transit trucks will deliver extended range, quicker refueling while emitting zero emissions. "Tim Slater, chair of Ford of Britain, said, 'Ford believes that the primary application of fuel cells could be in its largest, heaviest CVs to ensure they are emission-free, while satisfying the high daily energy requirements our customers demand." (Ford

Media, 2023) "Ford has researched fuel cell technology since the 1990s, developing many prototypes and demonstrating the first-generation E-Transit fuel cell vehicle at the CENEX Low Carbon Vehicle Show in 2021." (Ford Media, 2023) The Ford test fleet will run over a 6-month period for the next three years. "The prototype Ford E-Transits will be fitted with a high-power fuel stack, in conjunction with significant hydrogen storage capacity, optimized for safety, capacity, cost, and weight. An important project element will evaluate efficient and viable recycling for end-of-life components." (Ford Metra, 2023)

At the North American International Auto Show, Ford introduced a plug-in hybrid with a hydrogen fuel cell Airstream concept. The Airstream concept brings back the Airstream name from a 2007 concept. "The system, called HySeries Drive, is powered by a 336-volt lithium-ion battery pack at all times and has a range of 25 miles on full electric power. Once the battery pack is depleted by about 40 percent, the hydrogen-powered fuel cell begins generating electricity to recharge the batteries, increasing range another 280 miles, for a total driving range of more than 300 miles." (RP news wire, 2022) The HySeries Drive has fuel, engine, and energy-conversion flexibility. Ford is looking to patent the HySeries Drive technology.

Ford has introduced a plug-in hybrid with hydrogen fuel cell technology. This development is called the HySeriesDrive. The HySeriesDrive has engine flexibility. It could be an all electric lithium-ion battery which could be switched to hydrogen fuel cell system. Or it could be fuel cell, hydrogen tank that could be switched to a diesel tank and diesel engine.

General Motors

In 2020, General Motors announced that they are not planning any hydrogen fuel cell passenger vehicles in the near future. Instead, General Motors plans to introduce 20 new electric vehicle models in their showrooms in 2023. Although General Motors has current hydrogen powered passenger vehicles, they have a history of hydrogen fuel cell technology dating back to 1966.

In May of 2022, General Motors changed their plans for hydrogen fuel cell technology. General Motors has set up a startup company called Hydrotec, which will work on hydrogen fuel cell technology. The company is working on HFC for semis and trains and has established collaboration with some companies as Hydrotec continues its research into fuel cell technology. The fuel cells were developed in collaboration with Honda Motor Company. General Motors has announced its intent to have all electric cars in their showrooms. "The automaker foresees both technologies as avenues to a zero-emissions future as it continues to make internal-combustion vehicles for at least the next decade." (Hall, 2022) General Motors intends to work on a market for semi-trucks, buses and trains. "GM already has shown in its demos that a fuel cell system can power a normal consumer vehicle, but

the automaker recognizes fuel cells are better suited for larger applications."
(Hall, 2022) The Hydrotec lab is currently doing research and development,
pre-production testing for durability, single-cell testing, and fuel cell stacks.
"And what's developed and tested in this lab will go directly into what's built at
the manufacturing joint venture operation, Fuel Cell System Manufacturing
LLC, that GM operates with Honda in Brownstown Township." (Hall, 2022)
The plan is for the hydrogen fuel cell powered trucks to share the Chevy
Silverado platform.

"General Motors is the only company developing and commercializing
both hydrogen fuel cells and EV battery technology, and the Ultium Platform
and HYDROTEC fuel cell power cubes deliver where it matters most—performance and cost. This is opening new revenue potential for the company
as industries—including freight trucks, aerospace, aircraft, locomotive—turn
to GM for this technology to improve performance and reduce emissions."
(Hydrotec, 2022)

GM is looking to partner with Navistar to produce hydrogen fuel cell
applications in large trucks. "Hydrogen fuel cells offer great promise for
heavy-duty trucks in applications requiring a higher density of energy, fast
refueling and additional range," Navistar Chief Executive Persio Lisboa said
this week as he announced plans to bring the new International RH Series
of fuel cell trucks to market during the 2024 model-year." (Eisenstein, 2021)

Like Navistar, Nikola is also partnering with GM and hopes to have
Class 7 and Class 8 trucks with a range of as much as 1,200 miles.

1996 Electrovan: In 1966, General Motors modified a GMC van with a
hydrogen fuel cell. They named the hydrogen powered van the Electrovan.
The Electrovan was designed to match standard delivery van standards in
acceleration, performance and driving range. Pressurized tanks of hydrogen
and oxygen were installed in the van. The van was powered with an electric
motor. Floyd Wyczalek was the project manager of 206 employees working
on the first hydrogen fuel cell technology. " The Electrovan was considered

the most advanced electric vehicle ever built. As the first fuel cell powered automotive vehicle, it was rated as a major technical achievement in 1966." (GM Heritage, 2020)

2000 Precept: The Precept was a concept car powered by a diesel/electric hybrid vehicle. In 2000 the Precept was developed into a hydrogen fuel cell vehicle. The hydrogen fuel cell Precept had a range of 500 miles and could accelerate from 0 to 60 in 9 seconds. The Precept was created by Advanced Technology Vehicles group and the Global Alternative Propulsion Center.

2001 Opel Zafira-based HydroGen1: The Opel Zafira minivan was ca[able of 85 mph and set 15 world endurance records. The Opel Zafira was the bases for the HydroGen1.

2002 AUTOnomy: This was an exciting looking concept vehicle. Unlike other concepts that basically were regular vehicles with a hydrogen fuel vcell in place of the gasoline engine, this was an aerodynamic dramtically engineered experimental vehicle built from the ground up. "The AUTOnomy concept was the world's first vehicle designed completely around a fuel cell propulsion system. AUTOnomy was more than just a new concept car; it was hailed as potentially the start of a revolution in how automobiles are designed, built and used. AUTOnomy was the first vehicle to combine fuel cells with by-wire technology, which allows steering, braking and other vehicle systems to be controlled electronically rather than mechanically."(GM Heritage, 2020) "The fusion of fuel cells and by-wire technology opened the door to tremendous styling and design opportunities, every fuel cell vehicle shown so far has attempted to stuff the fuel cell stack, hydrogen storage unit and electric motors into existing internal combustion architecture, often at the expense of passenger space and payload capacity. But a fuel cell stack can be spread around the vehicle and can take any shape Bob might imagine. It doesn't have to be bunched up like the cylinders of an internal combustion engine."(GM Heritage, 2020)

2002 Hy-Wire: Unlike the very aerodynamic sports car appearance of the AUTOnomy, the Hy-Wire resembled a normal four door compact sedan. The HY-Wire was the first drivable vehicle that combined a hydrogen fuel-cell with by-wire technology. "GM developed Hy-Wire as a drivable concept vehicle in just eight months, showing a commitment to this technology and the speed at which the company was progressing. Hy-Wire accelerated this progress with a functional proof of concept which strengthened the company's confidence to gain marketplace acceptance of production fuel cell vehicles."(GM Heritage)

2004 HydroGen3: The HydroGen3 had more power, an easier start up procedure and a more compact fuel cell stack than the HydroGen1. A HydroGen3 was driven 6200 miles from Norway to Portugal across 14 countries over diverse road and climate conditions. "GM and FedEx launched the first commercial test fleet of fuel cell vehicles in Japan. HydroGen3's 250-mile range was the highest of any fuel cell vehicle approved for roads in Japan."(GM Heritage, 2020) "The U.S. Postal Service announced that it would lease a GM fuel cell powered minivan to deliver mail in the Washington, D.C. area, marking the first commercial application of any fuel cell vehicle in the U.S."(GM Heritage, 2020)

2005 Sequel: The Sequel had an appearance like most SUVs of its day. The Sequel was significant because virtually everything was based on the 11 inch skateboard chassis stemming from the AUTOnomy concept car. "The Sequel was the first vehicle in the world to successfully integrate a hydrogen fuel cell propulsion system with a broad menu of advanced technologies such as steer- and brake-by-wire controls, wheel hubs motors, lithium-ion batteries and a lightweight aluminum structure."(GM Heritage, 2020) "Sequel points to a vehicle that, in the future, will be better in nearly every way—quicker, surer-footed, easier to handle, easier to build, better looking, safer and only emits water vapor."(GM Heritage, 2020)

2007 "Project Driveway": In 2007 a fleet of 119 Chevrolet Equinox SUV's were driven daily over 3 million miles by more than 5,000 consumers in California, New York and Washington, D.C. This was the largest hydrogen fuel cell vehicle fleet ever assembled. It was one of the first meaningful marketing tests of a fuel cell powered vehicle.

2012 General Motors partnered with the U S Army to build a fleet of hydrogen fuel cell vehicles based on the Chevy Equinox crossover. The following year, **2013**, Gm and Honda announced a fuel cell collaboration with Honda to build a national hydrogen infrastructure and in **2017** they announced their intention to build a fuel cell factory in Ohio. In **2018** GM reaffirmed its commitment to include fuel cell vehicles in its lineup by 2020.

In **2020**, General Motors announced that they are not planning any hydrogen fuel cell passenger vehicles in the near future. This contradicts what they said in 2017. But in 2016, General Motors considered modifying the Chevrolet Colorado pickup truck for military purposes. "General Motors and the U.S. Army Tank Automotive Research, Development & Engineering Center (TARDEC) modified a Chevrolet Colorado midsize pickup truck to run on a commercial hydrogen fuel cell propulsion system and will expose the truck to the extremes of daily military use for 12 months."(GM Heritage, 2020) "Although General Motors has already gone on record in saying that a hydrogen powered consumer vehicle is currently off the docket, the automaker is nonetheless engaged in a strategic partnership with Honda to develop hydrogen technology, as well as manufacture fuel cells in Michigan. "GM Defense already has hydrogen fuel cell-powered vehicle concepts in the way of the Chevrolet Colorado ZH2 and Silverado ZH2. These emissions-free, four-wheel-drive fuel cell pickups were developed so the U.S. Army could determine the viability of hydrogen-powered vehicles on military missions. "Fuel cells have the potential to expand the capabilities of Army vehicles significantly through quiet operation, exportable power and solid torque performance, all advances that drove us to investigate this technology further, 'GM Defense said in a statement released back in 2016.'" (Mceachern, 2020) GM Defense is also working on an unmanned submarine for the army that is powered by a hydrogen fuel cell."(Mceachern, 2020) Now, recently comments from GM Defense President David Albriton suggest that a GM hydrogen fuel cell vehicle could be on the horizon for fleets—including military customers."(Lopez, 2020) "Hydrogen fuel cell technology is important to GM's advanced propulsion portfolio, and this enables us to put our technology to the test in a vehicle that will face punishing military duty cycles,' said Charlie Freese, executive director of GM's Global Fuel Cell Engineering activities."(GM Heritage, 2020) "There was no actual program of record within the U.S. Army, but we worked with the ground vehicle services center to test that vehicle and allow soldiers to actually get in it,' Albritton said. That was a great learning environment for us to actually learn how you can take a fuel cell and put it into an army vehicle with all the rigors of taking it off-road and everything else." (Lopez, 2020) "Albritton also discussed the Surus concept. For those that may be unaware, the Surus concept (Silent Utility Rover Universal Superstructure) was a GM hydrogen fuel cell platform geared for commercial and military applications, which provided General Motors with feedback on what the U.S. Military would require should something similar be adopted into service." (Lopez, 2020) Besides military application, GM also has worked with the U.S. Postal Service. "The U.S. Postal Service announced that it would lease a GM fuel cell-powered minivan to deliver mail in the Washington, D.C. area, marking the first commercial application of any fuel cell vehicle in the U.S." (GM Heritage Center, 2020)

In **2023**, Forbes reported that GM would use fuel-cell power for Navistar trucks. "General Motors as part of its plan to be a carbon-neutral company with an 'all-electric future' in 20 years, announced this week it will supply fuel cells to Navistar, a leading maker of big trucks." (Eisentein, 2023) "Hydrogen fuel cells offer great promise for heavy-duty trucks in applications requiring a higher density of energy, fast refueling and additional range,

'Navistar Chief Executive Persio Lisboa said this week as he announced plans to bring the new International RH Series of fuel-cell trucks to market during the 2024 model-year." (Eisenstein, 2023) With this announcement GM joins Honda, Hyundai, and Toyota in push for fuel cell technology.

Chrysler-Fiat

As early as 2001, Chrysler was working on a hydrogen fuel cell vehicle. In 2001, Chrysler introduced the Natrium, a fuel cell version of the Chrysler Town and Country minivan. The Natrium was shown at the 2001 Electric Vehicle Association of the Americas in Sacramento, California. Chrysler said they had been working on the Natrium for 18 months. The hydrogen is derived from sodium borohydride, a chemical that is similar to the laundry detergent, borax. One drawback was the need for filling stations to have two tanks, one for with fresh fuel and the other to dump the spent fuel. The Natrium never reached production. Sodium borohydride didn't pan out as a power source for hydrogen fuel cell vehicles.

Harold Wester, the Chief Technology Officer for Fiat-Chrysler said, "Hydrogen fuel cells are a better choice for future power trains than batteries." (Gerencars, 2020) Wester is also the CEO of Maserati and Alfa Romeo, said he was discussing possible alternative for these two vehicles as well. "Wester also claimed that continued use of fossil fuels to generate electricity ultimately makes battery-electric cars less green than fuel-cell vehicles." (Greencars, 2020)

Land Rover Jaguar

"As governments attempt to reduce emissions and boost urban air quality, the vehicles people use look set to change. Jaguar Land Rover is one of several automotive companies working on hydrogen powered vehicles." (Frangoul, 2021) In 2022, Land Rover announced that they are considering building a hydrogen fuel cell Defender. Little is known about the Defender other than testing began towards the end of 2021. "Hydrogen-powered FCEVs provide high energy density and rapid refueling, and minimal loss of range in low temperatures, making the technology ideal for larger, longer-range vehicles, or those operated in hot or cold environments, 'the company added." (Frangoul, 2021) "The FCEV is part of the company's strategy to achieve zero tailpipe emissions by 2036 and net zero carbon emissions across the supply chain and operations by 2039. Fuel cell electric vehicles generate electricity from hydrogen to power their electric motors, and Jaguar Land Rover considers them complementary to battery electric vehicles. 'We know hydrogen has a role to play in the future powertrain mix across the whole transport industry, and alongside battery electric vehicles, said Ralph Clague, head of the company's hydrogen and fuel cell division." (Lyons, 2021)

Jaguar Land Rover's Defender project is backed by Advanced Propulsion Centre and partly funded by the U.K. government. The project will use an existing Defender chassis rather than create an all-new body style. The testing will consider the Defender's attributes for off-road capability and fuel consumption. The project is dubbed," Project Zeus." The aim is to develop fuel cell vehicles that can deliver the same performance as the conventional Defender. Ralph Clague issued this statement in part. "The work done alongside our partners in Project Zeus will help us on our journey to become a net zero carbon business by 2039, as we prepare for the next generation of zero tailpipe emissions vehicles." (Moon, 2021)

Australia

One of the drawbacks to electric cars is that they cost too much. Australia is evidence that that statement holds true for hydrogen fuel cell cars and trucks as well. H2X Global produces a pickup truck named H2X Warrego, with a base price of $189,000. There are two other versions, Warrego 66 priced at $235,000 and the Warrego 90 priced at $250,000. Besides the Warrego ute,

H2X also produces the Darling, a van that can also be configured as a taxi. H2X also has trucks, the Renova and garbage trucks that are being used in Sweden. H2X CEO Brendan Norman anticipates the hydrogen fuel cell vehicle like the Darling playing an important role as a taxi. "Norman foresees that on a typical day, a hydrogen delivery van or taxi, would be used for a six-hour shift before refueling, which could be completed within three minutes. The vehicle could then be put back on the road with a driver change, essentially for permanent use." (Dorgan & Ogg, 2022) The refueling time is expected to be 3–5 minutes. "H2X Global says FCEV models can, in some areas, outperform battery electric vehicles (BEVs)—such as the speed of refueling times and what it says is 'limited expected life' for EV batteries, and 'issues with the disposal of lithium batteries." (Hill, 2021)

H2X was registered as a business in Australia in 2021, so it is a newcomer on the scene of hydrogen fuel cell vehicles. H2X CEO Brendan Norman, is not a newcomer having previously set up the Chinese HFC car called the Grove. "H2X has two operating divisions; one is designing and delivering the powertrain systems to heavy equipment and stationary power applications; the second develops and delivers multiple light equipment vehicles using a proprietary H2Xfuel cell and powertrain system." (Dorgan & Ogg, 2022) H2X has an aggressive marketing plan with plans to market not only in Australia but also Scandinavia, Europe, North Asia, and North America. "Under the agreement with the Gippland Circular Economy Precinct (GCEP), a consortium of businesses that recognize the need to transition the local economy from brown coal power generation to renewables, H2X will manufacture hydrogen fuel cells, electrolysers, hydrogen fuel cell-powered vehicles and a range of hydrogen power units including generators and emergency power supplies in the region."(Dorgan & Ogg, 2022) H2X plans to build manufacturing plants in India, Malaysia, Scandinavia and of course Australia. "The Sydney-headquartered company will provide hydrogen fuel cell-powered trucks and light vehicles to the city of Gothenburg, in a deal that H2X believes will pave the way for it to develop and produce vehicles for the wider Scandinavian transport industry in the future." (Dorgan, Ogg, 2022)

In 2022, Australia had four manufacturers offering hydrogen fuel cell vehicles available. Hyundai Nexi had a fleet of 20 available for the ACT government. Toyota Mirai was hoping that the infrastructure would catch up with the sale and leasing across Australia. Toyota had plans to build hydrogen stations across the country. H2X had pickup trucks available for purchase or lease. Ineos Grenadier an United Kingdom company in a deal with Hyundai a SUV Grenadier, which looks similar to the Land Rover Defender available to the public.

India

"Toyota Kirloskar Motor has unveiled a project to introduce India's first hydrogen=powered fuel cell electric car as one of the biggest emerging car markets to expedite its transition to clean transport." (livemint, 2022) The study will consider the hydrogen fuel cell car's ability to handle Indian roads and climate conditions. Toyota introduced the Mirai to India and issued this statement. "The statement had noted that green hydrogen offers huge opportunities to decarbonize a range of sectors, including road transportation, and is gaining unprecedented momentum globally."(Livemint, 2022) The study is being conducted along with International Center for Automotive Technology (ICAT).

China

Shanghi SinoFuel Cell hopes to leverage fuel cell vehicles over standard EVs, because of shorter charging times, longer cruising ranges, and the ability to perform at low temperatures. "China's Refire, which is second in the domestic market, has delivered fuel cell trucks to Swedish furniture chain IKEA and

49-tonne fuel cell trucks to U.S. construction machinery maker Caterpillar." (Doi, 2023)

"Nissan, traditionally associated with battery-powered electric vehicles, has entered the burgeoning hydrogen mobility landscape in China through its joint venture with DongFeng." (energynews, 2024) The collaborative effort has taken the existing Venucia vehicle and modified it with a hydrogen fuel cell and a lithium battery system. The Venucia H2e is reported to be able to run in extremely cold conditions. "The zero-emission vehicle is powered by H2 and a lithium-ion-phosphate (LFP) battery." (hydrogenfuelnews, 2023) The car is called the Venucia H2e. The "e" represents the battery portion. The Venucia is priced at $137,000 dollars but currently it is not for sale to the public. Instead, the DongFeng-Nissan vehicle has a fleet of cars in a test market for 36 months. In 2023 there was an uptick in hydrogen fuel cell vehicles, bhut they were mostly commercial sales. Despite skepticism hydrogen mobility continues to grow in China.

Another new car manufacturer to enter the hydrogen fuel cell market was Grove Hydrogen Automotive, a Chinese auto manufacturer. Grove was born in 2016, became registered in 2018. "Grove is a Global car company aiming to offer a truly clean Automotive experience from Manufacturing to the enjoyment of the car." (Chen, 2019) "Grove Hydrogen was established by the Chinese Institute of Geosciences and Environment, which currently manufactures and distributes hydrogen extracted from industrial waste. The institute says it is working with large Chinese cities to install and expand hydrogen charging infrastructure in the coming years."(Autocar Pro News, 2019) Grove has made a total commitment to fuel cells offer great promise for heavy duty trucks in applications requiring a higher density of energy, fast refueling and additional range, said Prsio Lisboa, Navistar president and CEO, 'We are excited to provide customers with added flexibility through a new hydrogen truck ecosystem that combines our vehicles with the hydrogen fuel cell technology of General Motors and the modular, mobile and scalable hydrogen production and fueling capabilities of OneH2. And we are pleased that our valued customer, J B Hunt has committed to utilize the solution on dedicated routes and to share key learnings." (Navistar, 2021) OneH2 makes hydrogen fueling stations and modular, mobile and scalable hydrogen production. "Navistar plans to make its first production model International RH Series fuel cell electric vehicle (FCEV) commercially available in model year 2024." (Navistar, 2021) OneH2 would supply hydrogen fueling solutions for Navistar which includes hydrogen production, storage, delivery and safety. Navistar believes its semitruck will have a range of 500 miles and could be refueled in 15 minutes. of Oneon sale next month in San Francisco and Los Angeles, can store more hydrogen than its predecessor, giving the sleek sedan a 30% increase in range." (Korn, 2020) components used in the production H2 Speed from Green GT include the 250kW fuel cell and torque vectoring system." (Smartcharger, 2020) The H2s supercar will only have a production run of 12 track only supercars . The cars are now market ready with a price tag of approximately 2.5 million dollars. There already is a list of prospective buyers.

Grove

Another new car manufacturer to enter the hydrogen fuel cell market was Grove Hydrogen Automotive, a Chinese auto manufacturer. Grove was born in 2016, became registered in 2018. "Grove is a Global car company aiming to offer a truly clean Automotive experience from Manufacturing to the enjoyment of the car."(Chen, 2019) "Grove Hydrogen was established by the Chinese Institute of Geosciences and Environment, which currently manufactures and distributes hydrogen extracted from industrial waste. The institute says it is working with large Chinese cities to install and expand hydrogen charging infrastructure in the coming years."(Autocar Pro News, 2019) Grove has made a total commitment to produce only hydrogen fuel cell vehicles. Grove claims to be the world's first hydrogen fuel cell only mass production company. The company is based in Wuhan, China, a city internationally known for another reason in 2020. Grove has entered into an agreement with designer Pininfarina to design its cars. The design center will be in Barcelona, Spain. Grove brought concept vehicles to the Shanghai Auto show including a 4 door sedan and a supercar sports car. Grove intends to produce a sedan, SUV and the supercar, all with the influence of designer Pininfarina. The four-door concept will have a bold grill and a highly sculpted rear diffuser and will feature frameless doors and cameras in place of mirrors. The four door SUV will be called the Obsidian and will have an estimated range of 625 miles. . The factory will be located in Chongqing, China, with sales starting in2019 and full volume production in 2020. Sales will concentrate on China with future expansion into Australia and New Zealand.

Grove has issued the following statement of philosophy: "Grove creates cars from the Earth and for the Earth, we respect the energy and materials the planet has given us. Born from the desire to create a car that is truly clean, from the earth it originates to the roads upon which it allows our passengers to roam, Grove has searched the earth to find the greatest scientific advances to achieve this Goal."(Grove Hydrogen Automotive, 2020) Grove says its cars will be environmentally friendly, have the advantage of clean energy, cost less on the road, require less hydrogen-fueling time and support a longer journey without refueling, compared to traditional vehicles. Grove expanded its philosophy with the following statement: "Not content with making a cleaner more efficient car at Grove, we focus aggressively on lowering our production driven emissions and also the use of precious resources on our planet. Our production process avoids the heavy welding and painting procedures of traditional cars, massively reducing the environmental impact in the manufacture of our cars."(Grove Hydrogen Automotive, 2019) "Grove Hydrogen Automotive Company Limited celebrated its first cars produced from its Nan'an District Chongqing production factory in Southwest China, April 13, 2019."(Chen, 2019)

The highlight of Grove's Shanghai Auto show exhibit was its supercar exhibit. This Pininfarina designed H2 Speeds, supercar followed the current state of the art supercar racers. Very aerodynamic. "The 503-HP H2 Speeds will use the chassis of a LeMans Prototype 2 racing model as a base and rely on a pair of electric motors that draw their power from hydrogen fuel cells for speed." (Bleier, 2019) The estimated top speed is 186 MPH and a zero to 60 time of 3.4 seconds. The car has been tracked tested by GreenGT, a Franco-Swiss company known for developing clean propulsion systems. "Twelve of these hydrogen race cars will be produced, powered by four electric motors that are good for a total output of 653 horsepower. Pinifarina's partner in the project is GreenGT, a Franco-Swiss company that designs and develops sustainable propulsion systems. Some of them produce only hydrogen fuel cell vehicles. Grove claims to be the only world's first hydrogen fuel cell mass production company. The company is based in Wuhan, China, a city internationally known for another reason in 2020. Grove has entered into an agreement with designer Pininfarina to design its cars. The design center will be in Barcelona, Spain. Grove brought concept vehicles to the Shanghai Auto show including a 4-door sedan and a super-car sports car. Grove intends of Oneon sale next month in San Francisco and Los Angeles, can store more hydrogen than its predecessor, giving the sleek sedan a 30% increase in range." (Korn, 2020) components used in the production H2 Speed from Green GT include the 250kW fuel cell and torque vectoring system." (Smartcharger, 2020) The H2s supercar will only have a production run of 12 track only supercars . The cars are now market ready with a price tag of approximately 2.5 million dollars. There already is a list of prospective buyers.

Russia

Russia has produced a hydrogen powered car called the Aurus, Senat. It comes in three body styles, a four-door sedan, a two-door convertible, and a four-door limousine. There is talk of a SUV, to be called the Komendant. but it has not surfaced. The Senat first appeared in 2013 and resembles a four-door Chrysler. The current Senat went into production in 2021. "The Russian premium car brand Aurus plans to sell its first hydrogen-powered sedan for $597,000." (Rallypulse, 2024) The most affordable version will sell for around $398,000. The Aurus, is capable of receiving electricity or hydrogen power.

"The car is equipped with a combined electrical installation, which includes a traction electric drive and an electrochemical hydrogen generation. The car will receive an all-wheel drive system, as well as a 100kWh battery, which was created especially for this car." (Hydrogen-Central, 2022) "Currently, only demonstration and test samples are being produced in order to optimize the technology, develop standards and amend the technical regulations of the Russian Federation in relation to electric and hydrogen vehicles." (Hydrogen-Central, 2022)

South Korea

"Hydrogen, as an energy source can be deployed across a wide range of uses— from power generation and transportation fuel to industrial manufacturing and heating buildings. Critically, using it a sa fuel generates no harmful emissions and the clean technologies to scale its production are available today. Underscoring this potential, the International Energy Agency predicts that hydrogen-could account for 10 per cent of global energy consumption by 2050." (Macquarie, 2022) "All parts of the hydrogen economy must simultaneously grow together." (Macquarie, 2022)

South Korea has set a goal of 300,000 Hydrogen car sales by 2030. However, Hydrogen vehicle sales dropped by 50% in 2023. Two factors were blamed for the drop, the lack of hydrogen charging stations and rising H2 fuel cost. That goal made need to be adjusted. Outside of China, South Korea has the largest FCEV market. In 2023, South Korean automaker Hyundai had more than half of the global hydrogen fuel-cell market. Worldwide hydrogen fuel-cell vehicle sales were up by 4.5 percent.

Japan

"The Japanese government has set an ambitious target of 800,000 fuel cell hydrogen cars on the nation's roads by the end of the decade." (Dokso, 2024) "Hydrogen car sales in Japan have fallen by 83% over the past two years, new figures show." (Collins, 2024) Has the Japanese government set too a high a goal? Almost ll of the fuel cell cars sold were by Toyota. Honda discontinued their hydrogen fuel cell car, the Clarity. When there was a small number of sales a drop of a few hundred could statistically look like a large percentage. A number that could be misleading. The future might be enhanced with various imported models like the South Korean Hyundai. "The primary goal of Japan's initiative to promote fuel cell hydrogen cars is to reduce greenhouse gas emissions and dependence on fossil fuels in the transportation sector. By encouraging the adoption of hydrogen fuel cell vehicles (FCEVs), the government aims to accelerate the transition to cleaner and more sustainable mobility solutions." (Dokso, 2024)

Trains

Trains might be the first form of transportation that will make the transition to hydrogen fuel cells. "Transitioning from diesel trains to hydrogen fuel cell electric trains is a promising way to decarbonize rail transport. That's because the fuel cell electric trains have several advantages over other electric trains, such as lower life-cycle emissions and shorter refueling time than battery ones, and less requirements for wayside infrastructure than the ones with overhead electric wires." (Ding & Wu, 2024) "Climate change and emissions reductions are topics high on the agenda for the rail industry. As companies continue to seek more sustainable fuel options, we explore the potential of hydrogen as a train fuel." (Youd 2021) "Fuel cell trains will play a key role in the transition to a zero-emission economy. Hydrogen powered trains are poised to disrupt the rail industry as a cost-effective, high performing, zero-emission alternative to diesel." (Ballard, 2024) A hydrogen train is one that is either powered by hydrogen or by a hydrogen fuel cell. Hydrogen fuel cell powered trains have produced new vocabulary. "Hydrail" is basically a train powered by hydrogen. "Hydrolley" is what we call a streetcar or trolley powered by hydrogen. The term "Hydrolley was coined at the Fourth International Hydrail Conference, Valencia, Spain, in 2008. Why hydrogen? "The use of hydrogen as an alternative rail fuel brings many potential benefits, the most noticeable being that is a clean energy source that supports zero-carbon strategies. Hydrogen fuel cell technology also provides a more powerful and efficient energy output compared to fossil fuels." (Youd 2021) Hydrogen fueling stations could be installed in the maintenance depots. In the past, the cost of rail service to rural areas was not feasible because of the cost of installing electric overhead lines. With hydrogen fuel cell trains that rural service could be made possible. Hydrogen fuel cell trains are a good alternative to diesel powered trains. Ballard Industries lists four important advantages of hydrogen fuel cell trains. 1. Fuel cell trains are as flexible and versatile as diesel powered trains. 2. There is no requirement for overhead catenary infrastructure. 3. They can be refueled in less than 20 minutes. 4. They can be gradually added so that the changeover can be aligned with the budget.

Trains are major carbon emitters. "Electric trains are common, but traditionally require costly catenary power lines or electrified rails to operate. By contrast, a fuel cell system creating electricity onboard the vehicle allows trains to run on existing tracks with no additional investment needed." (Ohnsman, 2019) Hydrogen fuel cell powered trains can have the same range as the current powered trains. Fuel cell stations can be installed right in the rail yards. Overall operating costs will drop with the use of hydrogen fuel cells. Canada has already put in strict carbon emissions requirements for its railway system. These Canadian regulations include emission standards for new locomotives, emissions testing, labeling and anti-idling requirements. The UK is about to test a new hydrogen fuel cell train called the Hydroflex.

"In the midst 0f the climate crisis, the demand for decarbonization across transport industries has grown and the Hydroflex is just one product of that. In 2016, Germany unveiled the Coradia iLint, the world's first hydrogen powered trains, which can run for 600 miles on a single tank of fuel—on par with the distances that traditional trains achieve on a tank of diesel. "In 2021, the Coradia ilint in Germany earned the European Railway Award as the first hydrogen passenger train in mainline operation." (Cummins, Inc. 2021) Engineers in the US are working on bringing a version of a 'hydrail' to the states. However, since rail is already among the lowest greenhouse gas emitters in transportation, it remains to be seen whether the value of a massive overhaul of rail systems will be worth it." (Hirschlag, 2020) The introduction of hydrogen powered trains in Germany has not resulted in large contract for new hydrogen powered trains.

"While supporters of hydrogen power for trains point to the fact that the only 'emission ' is clean water, this ignores the fact that hydrogen must be made somehow. Some hydrogen is available as a by-product of industrial activity, but nowhere near enough is available to power trains in most countries. The 'green' approach to making hydrogen uses electrolysis to split the hydrogen atoms from water, but this requires lots of electricity—around 3 times as much as a conventional electric train." (Fender, 2020) "There are benefits to passengers too. Hydrogen-powered trains, like electric trains, are also incredibly quiet compared to their diesel counterparts. And unlike electric trains they are more resilient to network-wide disruptions" (Hirschlag, 2020) Fuel cell electric trains have several advantages lower life cycle emissions and shorter refueling time. Hydrogen fuel cell trains are not only clean but require little wayside infrastructure. There are still some challenges for hydrogen fuel cell trains. "First, although many studies suggest hydrogen storage and usage are safe, a large part of the public is still concerned with it's safety. Second, effective onboard hydrogen storage has difficulties because hydrogen gas has much lower volumetric density than conventional fuels." (Ding, Wu, 2024) One solution is to mount the hydrogen tanks on the roof. A second solution is to separate the hydrogen tanks from the passengers with a separate compartment reminiscent of the old coal tenors on trains.

The fuel for the Hydrofluex is stored in four high pressure tanks. Currently the fuel and power packs are located on the roof of the cars, the aim is to relocate them underneath the rail cars. "The fuel tanks on the Hydroflex, for example, have to be small enough to fit in an ordinary car that can pass through Victorian-era railway tunnels." (Hirschlag, 2020) "The way hydrogen powers a train like the Hydroflex is quite simple. The fuel cell is made up of an anode, a cathode and an electrolyte membrane. The stored hydrogen passes through the anode, where it is split into electrons and protons. The electrons are then forced through a circuit that generates an electric charge that can be stored in lithium batteries or sent directly to the train's electric motor. The left-over part of the hydrogen molecule reacts with oxygen at the cathode and becomes the waste product—water." (Hirschlag, 2020)

"Future critical developments needed to move the technology forward include higher efficiency fuel cell systems, taking advantage of lower projected costs and modularity, higher durability membrane electrode assemblies using advanced materials, tighter system controls and optimized operating conditions, and the ability to deliver hydrogen to the locomotives at competitive cost." (Ohnsman, 2019) "Hydrogen trains have already been put into operation, using hydrogen fuel cells as ongoing research investigates the use of converted hydrogen internal combustion engines." (TWI, 2022) "Regardless, hydrogen train technology could prove to be an important part of a wider transportation network, including trains and other local travel, passenger trains, goods transporting and industrial railways." (TWI, 2022)

At the time of this writing in 2023, here is a summary of HFC Train Manufacturers and programs that were in effect or in the planning stages. Alstom, a French company, produces the Coradia iLint the world's first HFC powered passenger train. The iLint was working or planned for in Germany, Austria, Poland, The Netherlands, United Kingdom and Italy. The Stadler FLIRT Arrow was planned to be in Switzerland by 2024. Siemens Mobility

Plus H in Germany produced the Mireo Plus H train, which is in the testing phase in Bavaria. The Mireo Plus H is also being tested in Talgo, Spain. In the U.K., the University of Birmingham Centre for Railway Research and Education in partnership with Porterbrook, developed the HydroFLEX. The HydroFLEX began operating in the United Kingdom in 2020. In the United Sates, California based Sierra Northern was converting a RailPower diesel into a HFC-powered switcher, with partial funding from the California Energy Commission. Also in the U S, BNSF, Caterpillar, and Chevron teamed up on a locomotive pilot pn the feasibility of hydrogen fuel as a viable alternative. And finally, in China, CRRC (China Railway Rolling Stock Corporation) developed a switching HFC locomotive.

"Russian Railways, along with Rusmano and the Sinatra Group, has agreed to work for the development of hydrogen fuel cell mainline locomotives." (Railways technology, 2021) Sinara will serve as the general contractor. Rusmano will be responsible for the development of hydrogen fuel cells, energy storage and other technological solutions. Russian Railways has issued the following. "From 2025, we plan to purchase only electric locomotives, as well as locomotives operating on alternative energy sources such as natural gas and hydrogen. This will enable us to further reduce the environmental burden on the environment." (Railway technology, 2021) Russian Railways intended to produce prototypes of both mainline and shunting locomotives by 2023–2024.

Buses

The first inroad to hydrogen fuel cell propelled vehicles is likely to have been hydrogen fuel cell buses. Buses are the most thoroughly tested areas of application of hydrogen fuel cell powered vehicles. A fuel cell bus is a bus that uses a hydrogen fuel cell as its power source for electrically driven wheels. "Unlike buses that run on fossil fuels, a hydrogen fuel-cell electric bus is powered by two of earth's most common basic components—oxygen and hydrogen. A fuel cell combines hydrogen and oxygen to produce electricity, heat and water. Fuel cells are similar to batteries. Both convert the energy produced by chemical reaction into usable electric power. The fuel cell provides an advantage, however—it will continue to produce electricity as long as fuel (hydrogen) is supplied." (OCTA, 2024) Modern fuel cell buses draw their energy from two fuel cell stacks. In 1998, there was a three-year trial in Chicago and Vancouver of hydrogen-powered fuel cell buses. The buses had New Flyer Industries bodies and Ballard Power System hydrogen fuel cells. In 2000, Hino and Toyota collaborated on the development of hydrogen fuel cell buses, the FCHV-BUS was later used in the Expo 2005.In 2001, London, Madrid and Hamburg began using hydrogen fuel cell powered buses. In 2006, the Federal Transit Administration issued a 49million dollar grant for the development and testing of hydrogen fuel cell buses. In 2009, a fleet of 20 hydrogen fuel cell b uses were used in British Columbia. In 2010, a fleet of 8 fuel cell buses were put into service in London. In 2015, Toyota began testing their hydrogen fuel cell buses in Tokyo. In 2018, 74 hydrogen fuel cell buses were put into service in preparation for the Beijing 2022 Winter Olympics. During the Olympics 800 hydrogen fuel cell buses were in operation. Since then the use of hydrogen fuel cell buses has boomed all over the world. By 2024, there were 14 different companies working on hydrogen fuel cell bus development. Mercedes-Benz believes that mass production lies in commercial vehicles, such as buses. "Public transit leaders know that zero-emission buses are critical to the future of transit. Many regions around the world are setting goals and mandates to convert entire fleets to electric buses." (Ballard 2022) In 1998, there were two-year three bus Ballard demonstration projects. Small urban fleets are being promoted as a way to provide technological development and add to the clean air efforts. From 2004–2006, another two-year, three bus trial took place in Oakland, California. AC Transit operated three buses and Sunline Transit Agency from Palm Springs; California operated one. The initial project was promising, and Santa Clara Valley Transportation Authority operated three more buses.

In the year 2006 in Beijing, China began operating hydrogen fuel cell buses on an experimental basis. In the same year, Daimler began operating three hydrogen fuel cell buses experimentally. In 2009, AC Transit operated 12 hydrogen fuel cell buses, featuring buses designed by Van Hool of Belgium. In 2009, the first Brazilian hydrogen fuel cell bus prototype began operation. Whistler, British Columbia had a fleet of hydrogen fuel cell buses for the 2010 Winter Olympics, but because of cost the program was halted in 2015.

In 2016, Orange County, California added its first fuel cell bus to its fleet and anticipated adding more in the future. A cleaner transit future is well under way at OCTA. We've just debuted a new hydrogen fueling station, the largest transit-operated fueling station in the United States, and are introducing 10 hydrogen fuel-cell electric buses. This $22.9 million program follows a pilot program in which OCTA became the first large public transportation agency in Southern California to operate a hydrogen fuel-cell electric bus. This program is designed to provide a clear path for protecting the environment." (OCTA, 2024) Japan had FCHV-BUS manufactured by Hino Motors and Toyota Motors. 2003–2004 there was a one bus demonstration by Toei Bus in Tokyo. In Expo 2005 Aichi, ran a 6-month trial with eight fleets of buses, that were thought to have carried over one million visitors. "Fuel cells improve the performance of electric buses by generating onboard power from hydrogen to recharge the batteries. Today, bus manufacturers offer fuel cell buses to transit agencies as a standard electric propulsion option." (Ballard, 2020)

With Whistler, British Columbia cancelling its hydrogen fuel cell program in 2015 and the City of Montpellier cancelling an order for 50 hydrogen fuel cell buses, the question arises, will hydrogen or electric buses win? Both Whistler and Montpelier said the program was too costly. Officials at Montpellier said that the hydrogen fuel cell buses were 6 times more expensive than electric buses. Claims were made that even in its most efficient form hydrogen production and distribution was way less efficient than electric battery powered buses. Jocelyn Timberley, writing in Science, pointed out that as the size of a vehicle increases, hydrogen becomes an increasingly more attractive option. (Timberley, 2021) "Battery electric and hydrogen fuel cell vehicles have similar propulsion systems. Both store energy to power an electric motor. However, in the latter, energy stored as hydrogen is converted to electricity by the fuel cell, rather than being stored in a rechargeable battery." (Timberley, 2021) As vehicles get bigger, it gets harder to electrify them, with increasingly larger batteries. (Timberley, 2021) A hydrogen fuel-cell bus may cost up to $900,000 in contrast to an electric bus might run from $500,000 to $800,000. "Hydrogen also has a relatively low density (the amount of energy that can be stored per unit volume mass area) around four to five times lower than petroleum fuels, but far higher than electric batteries, he adds." (Timberley, 2021) "Fuel cells improve the performance of electric buses by generating onboard power from hydrogen to recharge the batteries. Today, bus manufacturers offer fuel cell buses to transit agencies as a standard electric propulsion option." (Ballard, 2022) Hydrogen fuel cell proponents have argued that hydrogen technology will prevail. It is estimated that China has already over 5,000 hydrogen fuel cell buses in operation. Mark Teevan, HMI Chairman of Toyota Ireland had this to say regarding the 2030 target for zero-emission systems. "We are very conscious of the environmental challenge we face in meeting our 2030 targets and the need to find zero-emissions solutions that will satisfy the varying needs of different users. Public transport, haulage, van delivery, taxi or private car." (Irish Tech News, 2020)

Infrastructure is a problem for both hydrogen and electric vehicles. In 2022, the United States passed a bill to build a nationwide infrastructure for electric battery powered cars for refueling. For a fleet of buses, building a hydrogen refueling station at the bus terminal is all that is necessary for a bus fleet. Hydrogen fuel cells in cars can be refueled in 3 to 6 minutes. Far faster the electric car batteries. Hydrogen fuel cell vehicles have a longer range on a fill up than the electric battery-operated cars. The larger the vehicle the greater the advantage in range for the hydrogen fuel cell vehicles. Hydrogen fuel cell vehicles have an advantage over battery vehicles when the roads are hilly, and the environment tends to be hot and humid. The danger of hydrogen explosions is a problem, but manufacturers now say they have overcome the problem of hydrogen storage and the possibility of a hydrogen explosion has been overcome. The electricity for electric cars comes from fossil fuels, where hydrogen does not need mining or fossil fuels for its production. "For all the potential of hydrogen buses, their deficiencies may ultimately outweigh their benefits." (Timberley, 2021)

Ballard industries offer the following advantages for fuel cell electric buses.

- Zero-emission at the tailpipe

- Range of up to 300 miles to 450 km between fueling

- Can be refueled in less than 30 minutes.

- Consistent power delivery during the duty cycle, in heat and cold

- Compact depot gas refueling, eliminating the need for roadside charging infrastructure

- Proven durability, with fuel cell lifetime of more than 30,000 hours

(Ballard, 2022)

Ballard announces in August of 2023 that it had received orders for Solaris for almost 100 fuel-cell system for buses in Europe. Ballard was accelerating their zero-emission fleet through U.S. Department of Energy's office of clean energy development (OCED) recent funding.

9.

WHAT INFRASTRUCTURE
IS NEEDED?

TO ANSWER THE AGE-OLD QUESTION; "WHICH came first, the chicken or the egg? Do you build an infrastructure of hydrogen fueling stations along our Interstate highway system or do you build the hydrogen fuel cell cars first? Or taking a page from the baseball story, "Build it and they will come." build the infrastructure and the hydrogen fuel cell cars will come in great numbers. This requires dedicated hydrogen fueling stations and dedicated hydrogen powered cars. One possible solution would involve cooperation with current private gas stations. If current gas stations would convert just one of their pumps into a hydrogen fuel pump the infrastructure could quickly be built. This would require the cooperation of the present fossil fuel suppliers. Perhaps there could be some government assistance to the fuel companies. Production of hydrogen-fueled cars is limited because people won't buy those cars if hydrogen refueling stations are not easily accessible, and companies won't build refueling stations if they don't have customers with hydrogen-fueled vehicles. "While hydrogen is seen as a sort of happy medium between gasoline/diesel and pure electric, adoption of the technology has been slow due to the limited refueling infrastructure. A fleet hydrogen vehicle, say a truck or work van, could potentially sidestep this issue as some companies with large fleets also have their own refueling stations on site." (Mceachern, 2020) Regarding the infrastructure in California, Keith Malone of the California Fuel Cell Partnership offered this thought. "If we can build the stations, we can sell the cars, Keith Malone of the California Fuel Cell Partnership, an industry-government collaboration founded in 1999 to expand the domestic market, told ABC News. In the United States, as of 2020, about 60 hydrogen refueling stations for vehicles are operating. About 40 of these stations are available for public use, nearly all of which are in California. The State of California has a program to help fund the development of publicly accessible hydrogen refueling stations throughout California to promote a consumer market for zero-emission fuel cell vehicles. The need for comprehensive hydrogen fueling stations will grow as the market grows.

Mobile hydrogen fuelers, where liquefied or compressed hydrogen and dispensing equipment is stored onboard a trailer, are also being developed to support the expansion of hydrogen infrastructure.

"The hydrogen fuel cell technology depends firstly on having a sufficient amount of hydrogen produced using green energy sources and at competitive prices. Furthermore, the needed refueling infrastructure is currently lacking worldwide." (Dorofte, 2020) Once hydrogen has been produced and stored it needs to be distributed to the hydrogen fueling stations. This can be done by developing an infrastructure. There are a variety of ways to make his distribution. "Unfortunately, it's not as simple as transporting gasoline in a big tanker on the road (or sea), due to the high pressure the hydrogen is under, and increased flammability that hydrogen has (compared to gasoline). This makes transporting hydrogen more expensive especially over longer distances." (Perry, 2018) In 2020, Shell, Toyota and Honda announced plans to expand the hydrogen refueling network in California. In 2015, it was estimated that it would cost about $1.5 million per station. In 2017 there were about 30 refueling stations in California. "If successful, Shell Hydrogen will install hydrogen refueling equipment at 48 existing Shell retail stations, upgrade two current Shell Hydrogen stations and add light-duty fueling dispensers and position at one existing Shell Hydrogen heavy-duty truck station." (Telematics News, 2020) The Orange County Transit Authority, OCTA, Has announced a new fueling station. This will be the largest capacity fueling station in the nation with a capacity of 18,000 gallons of hydrogen fuel, which is enough to fuel 40–50 buses.

"The future is now, 'said Scott Nargar, Hyundai's senior manager for future mobility and government relations. 'We really need infrastructure on the forecourts of service stations, we need infrastructure in shopping centres, we need infrastructure in the home." Scott Nargar was speaking in 2019 on the need for infrastructure in Australia with the arrival of hydrogen fuel-cell vehicles in Australia.

In 2020, Airbus introduced three hydrogen fueled concepts for aircraft. "The success of such a program would depend on infrastructure at airports and support governments to fund development, as well as incentives for airlines to retire older aircraft, Airbus said. The company has already started discussions with airports, airlines, and energy companies. It is also calling on governments to put the right incentives in place to push the industry to shift toward hydrogen power." (Ryan, 2020)

Hydrogen Storage

Hydrogen storage has become a key factor in the advancement of hydrogen energy applications. "Hydrogen storage is a key enabling technology for the advancement of hydrogen and fuel cell technologies in applications including stationary power, portable power, and transportation." (eere, 2024) Hydrogen can be stored physically as either a gas or a liquid. "As an environmentally safe energy source, hydrogen has the potential to be cost-effective option for meeting rising energy demands. However, the three different forms of hydrogen, including compressed gas, cryogenic liquid, and solid-state form, provide significant storage and transportation difficulties." (Ahmed, 2021)

Compressed gas and liquid hydrogen storage need to be handled carefully. "There are several existing hydrogen storage methods, including compressed gas, cyro-compressed gas, liquid, physical absorbent, metal hydrides, and chemical hydrogen carriers." (Ding & Wu, 2024) Hydrogen can be stored in chemicals such as ammonia and methanol. "The ideal storage medium should allow high volumetric and gravimetric energy densities, quick uptake and release of fuel, operation at room temperatures and atmospheric pressure, safe use, and balanced cost-effectiveness. While not perfect, the current leading industry standard of compressed hydrogen offers a functional solution and demonstrates a storage option for mobility compared to other technologies." (Rivard, Trudeau, and Zaghib, 2019)

There are safety issues with hydrogen storage. The small size of hydrogen molecules allows them to penetrate through the storage vessel. One possible remedy is to wrap the hydrogen storage vessels in carbon fibre. Another possible remedy is to coat the storage vessel with cadmium-titanium or cadmium-nickel coating. "All current hydrogen storage technologies have significant drawbacks, including complex thermal management systems, boil-off, poor efficiency, expensive catalysts, stability issues, slow response rates, high operating pressures, low energy densities, and risks of violent and uncontrolled spontaneous reactions." (Rivard, Trudeau and Zaghib, 2019)

Liquid Hydrogen storage

"The hitch is that while an excellent medium for renewable energy storage, hydrogen itself is hard to store." (Willige, 2022) A liquid hydrogen tank is designed to hold a cryogenic liquid. It is not designed to withstand internal pressure. To handle the internal pressure, it has been discovered that the sphere is the best shape for a hydrogen storage vessel. The spherical shape also allows for less likely leaks. In time the landscape will be punctuated with two story high white sphere of liquid hydrogen storage tanks. There is even talk of future mega size storage tanks. "Heat transfer from the environment to the liquid increases the pressure inside the tank. Since the tank is not designed to hold high pressure, hydrogen is allowed to escape through a relief valve, which is sometimes referred to as 'boil-off.'" (Rivard, Trudeau and Zaghib, 2019) Liquid storage tanks will have a double wall vacuum insulation construction that will serve as insulation and a safety barrier should the inside tank develop a leak. This same vacuum insulation, double wall construction will also be used in the transfer lines connected to the storage tanks.

Cold/Cryo Compression Storage

Hydrogen can be compressed and stored in tanks and then used as needed. The volume of hydrogen is four times the volume of natural gas. For practical handling purposes hydrogen needs to be compressed. The hydrogen used in hydrogen fuel call vehicles is compressed. "Cryomotive's CRYOGAS solution claims the highest storage density, lowest refueling cost and widest operating range without H2 losses while using one-fifth the carbon fiber required in compressed gas tanks." (Gardiner, 2023) Cryogas is cold H2 gas stored in an insulated pressure vessel. The high density of Cryogas enables more H2 fuel in the tank for a longer period. Cryomotive is targeting heavy trucks for its first commercial application by 2025.

Metal-Organic Storage

Porous coordination polymers, also called metal-organic frameworks, (MOFs) are among the most challenging materials for hydrogen storage. Hydrogen gas can be stored by metal-organic frameworks. "Metal-organic frameworks (MOFs) are crystalline materials that have structural versatility, high porosity and surface area, which can adsorb hydrogen efficiently." Shet, et al, 2021) MOFs are currently in the trial-and-error research stage. "MOFs have a large specific surface area, and the adsorption site above it can be used for the adsorption and release of hydrogen (weak dispersion interaction)." (Zhang, et al, 2023) "MOFs can be used as an alternative to clean energy storage. The cost of fuel for hydrogen cars can be reduced by using MOF as

it seemed to have more hydrogen capacity than compressed hydrogen gas."
(Shet, et al, 2021)

Carbon Nanostructures Storage

The first experimental evidence for carbon nanotubes as a possible hydrogen storage vessel occurred in 1986 and later in 1991, carbon nanotubes (CNTs) became regarded as a good candidate for material hydrogen storage. Carbon nanotubes have been shown to store considerable amounts of hydrogen.

Metal Hydrides Storage

Metal hydrides are compounds containing metals and hydrogen. One of the most common examples is Magnesium hydride, attractive for hydrogen storage because of its abundance and affordability. Metal hydrides can be used in a variety of applications, including neutron moderation, electrochemical cycling, thermal storage, heat pumps, and purification/separation operations. The main advantage of hydrogen storage in metal hydrides for stationary application are the high volumetric energy, density and lower operating pressure compared to gaseous hydrogen storage. High temperature, high energy and slow kinetics are a problem for reversible storage. Complex hydrides operate over a broad range of temperatures. Since the hydrogenation and dehydrogenation reactions are endothermic and exothermic, it is unlikely that the heat generated or required by the chemical reactions can be kept in a closed loop or recycled. The safety of hydrides is also questionable." (Rivard, Trudeau, and Zaghib, 2019) Therefore, metal hydride storage would not be applicable to hydrogen fuel cell vehicles.

Metal Borohydrides Storage

Borohydrides find wide use as reducing agents in organic synthesis. The most important borohydrides are lithium borohydride and sodium borohydride. Lithium borohydride is an effective additive in enhancing desorption and absorption properties. Alkali metal borohydrides were first developed in 1940 by Hermann Irving Schlesinger and Herbert C. Brown.

Hydrogen storage has become a key to the application and development of hydrogen. "At present, there are three main ways of hydrogen storage: gaseous hydrogen storage, liquid hydrogen storage and solid hydrogen storage. Among them, solid-state hydrogen storage is a technology that stores hydrogen in solid hydride materials." (Front chem, 2022) "Solid-state hydrogen storage method has a very high volumetric hydrogen density compared to the traditional compressed hydrogen method. The main issue of solid-state hydrogen storage method is the development of advanced hydrogen storage materials. Metal borohydrides have very high hydrogen density and have received much attention over the past two decades." (Front Chem, 2022) In chemical hydrogen storage hydrogen is combined with various elements to create metal hydrides such as lithium borohydride. "However, metal borohydrides have high thermal stability, and generate highly inert element boron after hydrogen releasing, which affects its reverse reaction to absorb hydrogen again. Therefore, improving the reversible hydrogen storage properties of metal borohydrides has become one of the hotspots for solid-state hydrogen storage materials." (Front chem, 2022)

Chemical Hydrogen Storage

"It is evident that pure hydrogen is difficult to transport due to its physicochemical properties, such as low energy density. Therefore, it is worth investigating pathways to chemically store hydrogen. The term chemical hydrogen is used to describe the strategy of storing hydrogen by synthesizing molecules thar contain hydrogen, Methane, the simplest hydrocarbon, can be synthesized by a process k own as methanation." (Rivard, Trudeau, and Zaghib, 2019) "A common way to produce hydrogen from methane is steam reforming, but this reaction is highly endothermic, i.e., it requires a lot of energy. Consequently, it is not suitable for mobile applications." (Rivard, Trudeau, and Zaghib, 2019

Hydrogen Highways

Key challenges to hydrogen delivery include reducing cost, increasing energy efficiency, maintaining hydrogen purity, and minimizing hydrogen leakage. Delivery technology for hydrogen infrastructure is currently available commercially, and several U.S. companies deliver bulk hydrogen today. "With President Bush's State of the Union Address on January 30 calling once again for more federal money to be put into the advancement of hydrogen cars and vehicles, the idea that there will indeed be a hydrogen highway one day leaves no doubts." (Kantola, 2006) Growth in hydrogen demand will require regional expansion of this infrastructure and development of new technologies, such as chemical carriers to transport hydrogen at high density and high throughput fueling technologies for heavy-duty fuel cell transportation. Hydrogen is transported from the point of production to the point of use via pipeline and over the road in cryogenic liquid tanker trucks or gaseous tube trailers. Liquefaction plants, liquid tankers, and tube trailers are deployed in regions where demand is at a smaller scale or emerging.

Hydrogen Stations

Before hydrogen fuel celled vehicles will begin to make inroads into the market there needs to be a network of fuel stations. Hydrogen refueling stations (HRS) can be onsite storage tanks or directly supplied from a hydrogen pipeline. The fuel cell pumps will be similar in appearance to regular gasoline pumps with nozzles, keypad, and information screen. The major difference will be in the design of the nozzle. Hydrogen nozzles will have a barrel that fits over the fueling receptacle whereas the gasoline nozzle fits into the receptacle. "Fueling a hydrogen car comes naturally overtime, but aligning the heavy nozzle and sealing it properly so the car and pump can communicate electronically can require some practice." (Voelcker, 2022) The hydrogen fuel control computer disables the vehicle as a safety measure, allowing the vehicle to fill up safely. The fuel up takes only 5–10 minutes. A tremendous advantage of electrically charged vehicles. When the fueling is completed the driver pulls up on the handgrip latch to unlock the nozzle and return the nozzle to its holder. Hydrogen fueling stations will cost an estimated 1–2 million dollars apiece. Building these stations means the companies building them will have to overcome permits and approvals to build. When introducing its BMW NEXT in 2020, BMW noted the need for an extensive Europe-wide network of hydrogen fueling stations. "The rollout of heavy-duty hydrogen trucks such as line-haul trucks will necessitate very large stations compared to light-duty needs. At the point of hydrogen use, additional infrastructure components that are commonly employed include compression, storage, dispensers, meters, and contaminant detection and purification technologies.

In 2020 Hyperion introduced the prototype of the Hyperion XP-1 super car with production aimed for 2022. Angelo Kafantaris, the CEO of Hyperion wants to make an infrastructure of hydrogen refueling stations in advance of production. Because of the long range of the Hyperion, over 1,000 miles, fewer refueling stations will need to be built. "Hyperion has announced the introduction of so-called Hyper: Fuel Mobile Stations. The mobile energy stations contain a hydrogen dispenser for fuel cell vehicles and an optional DC fast charger for battery vehicles. The stations will be manufactured at the company's 65-acre manufacturing headquarters in Columbus, Ohio, according to the company, Hyperion's goal is to install them throughout the US and form a dynamic "Hyper: Fuel Network" with partners starting in 2023." (Randall, 2022)

The development of hydrogen fuel cell infrastructure is taking place more in Europe and Japan than in the United States. Japan has been more receptive to hydrogen fuel cell technology. One trend has been private development infrastructure than government efforts. BMW is committed to pushing forward the creation of hydrogen fuel stations throughout Europe. In 2015, Honda, Nissan and Toyota announced plans to cooperate in building infrastructure in Japan. "To promote the advancement of fuel-cell vehicles, Honda, Nissan, and

Toyota will collaborate on strengthening the hydrogen fueling infrastructure in Japan. The automakers have not yet announced specific plans for the project but said they would look at taking on a portion of expenses involved in operating hydrogen stations to help the government and infrastructure companies expand hydrogen fueling efforts. The automakers are working to achieve goals outline last year by the Japanese government, which will offer subsidies for installing and operating hydrogen fuel stations. All three automakers have been involved in fuel-cell technology for some time. Honda first introduced the FCX Clarity to lessees for the 2008 model year, before introducing a new fuel-cell concept at the 2013 Los Angeles auto show. Meanwhile, Toyota just debuted the 2016 Mirai that will arrive in showrooms this fall. Nissan brought out a Terra FCEV concept at the 2012 Paris motor show, but we've yet to see a production model from the automaker come our way."(Pleskot, 2015)

In November Hyundai announced a long-term plan to develop a reliable source of hydrogen in Europe. Hyundai and petrochemical giant Ineos are exploring opportunities to try to make hydrogen-powered happen on a larger scale than it has in 2020. Ineos has launched a new program to build hydrogen fuel cell capacity in Europe preparing for its own HFCV, the Grenadier. One way to overcome obstacles for hydrogen fuel cell infrastructure is for private enterprises like Hyundai and Ineos to build their own hydrogen fuel cell stations so that the purchase of their HFCV's will be more acceptable. "But if a vast network of hydrogen fuel stations were built across Europe, Ineos would be one company that could make it happen."(Ramey, 2020)

Hydrogenics produces a fueling station called HySTAT, which uses advanced electrolysis technology to split water into hydrogen and oxygen, using only electricity. "The HySTAT fueling station is a turnkey solution, which comes fully interconnected, automated and is easy to install. The process is quiet, reliable and safe; and it provides zero emission fuel from production to consumption, eliminating the dependency on carbon fuels." (Hydrogenics, 2020)

"Hydrogen infrastructure is also developing for buses, medium-duty fleets, and material handling equipment. Unlike the public, consumer stations for FCEVs that need multiple locations to cover wherever the consumer may travel, private, fleet fueling stations require fewer locations or even just a central location to meet a specific fleet's needs." (U.S. Department of Energy, 2020)

A major concern about hydrogen fueling stations is their safety. The fear of a station exploding is always present. In June of 2019, a hydrogen refueling station in Norway exploded causing the company opereating the station to suspend operations at its other locations following the explosion. Toyota and Hyundai both halted the sale of their hydrogen fuel cell cars.

"Jon Andre Lokke, CEO of Nel Hydrogen, the company operating those hydrogen refueling stations commented: 'It is too early to speculate on the cause and what has gone wrong. Our top priority is the safe operation of the stations we have delivered. As a precaution, we have temporarily put ten other stations in standby mode in anticipation of more informatio0n.'" (Lambert, 2019)

"Toyota Norway manager Espen Olsen said: 'We don't know exactly what happened on the Uno-X drive ye, so we don't want to speculate. But we stop the sale until we have learned what has happened, and for practical reasons, since it is not possible to fill fuel now.'" (Lambert, 2019)

The explosion raised the question as to the future of hydrogen fuel celled cars. Toyota issued this statement. "This does not change our view of hydrogen, and it is important for us to point out that hydrogen cars are at least as safe as ordinary cars. The hydrogen tanks themselves are so robust that Bob can shoot them with a gun without knocking them." (Lambert, 2019)

Hydrogen Pipelines

Using pipelines is a viable option, but it is limited by the lack of infrastructure. Currently America has about 1600 miles of pipeline, mostly in southern

California. Hydrogen can be stored in a liquid form which would allow for it to be transported because it is denser. The hydrogen can be liquified by a process called cryogenic liquefaction.

Under President Joe Biden fossil fuels, including natural gas, have come under attack. Natural gas pipeline owners are looking for alternatives for their pipeline business. Gaseous hydrogen can be transported through pipelines much the way natural gas is today. Hydrogen needs a pipeline infrastructure to make hydrogen fuel available. Transporting gaseous hydrogen via existing pipelines is a low-cost option for delivering large volumes of hydrogen. To this end, the gas pipeline provides the infrastructure needed. It is possible to send hydrogen and natural gas through the pipeline together and then separate them at the destination. Coverting natural gas pipelines to carry a blend of natural gas and hydrogen (up to about 15% hydrogen) may require only modest modifications to the pipeline. Converting existing natural gas pipelines to deliver pure hydrogen may require more substantial modifications. "The National Renewable Energy Lab (NREL) recently reviewed the challenges and opportunities of blending hydrogen gas with natural gas for transportation in existing steel pipelines. Such a practice could improve the environmental footprint of natural gas consumption, provide a way to shuttle hydrogen gas across the country, and be significantly less expensive than building out dedicated hydrogen pipeline networks." (Chatsko, 2020) At the present time there are about three million miles of natural gas pipeline in the United States. Pumping hydrogen could be a life saver for the natural gas pipeline owners. Unfortunately switching from one to the other presents contract problems, delivery problems and retooling infrastructure costs.

"Unlike natural gas, hydrogen can be burned without pumping carbon dioxide into the air. Run it though a fuel cell to generate electricity and the only waste is water; produce hydrogen using electrolyzers powered by solar plants or wind farms and it becomes a way to store massive amounts of renewable energy—far more than any of today's batteries can hold. And the best part for pipeline companies getting it where it needs to be, in bulk, could require the same basic infrastructure that now carries natural gas."(Baker, et al, 2021) An electrolyzer is a device that produces hydrogen. Presently there is more activity in the UK and Europe to build hydrogen infrastructure than there is in the United States.

Hydrogen Fuel Production Plants

While gasification technologies have advanced over the past several decades, gasification systems costs remain high. Historically, "economies of scale" have lowered prices, but the huge capital investment required for large plants and the accompanying financial risk have become a significant barrier to market penetration. Modular gasification-based energy conversion plants that are flexibly right-sized, configured, and sited to take advantage of local labor pools and utilize feedstocks of low cost coal, waste coal, coal fines, biomass, MSW, and waste plastics could be optimized to supply local and niche markets with power, combined heat and power, and fuels production, thereby conferring significant site-specific impacts and benefits. Gasification plants for biofuels are being built and perated, and can provide best practices and lessons leasrned for hydrogen production. The U. S. Department of Energy anticipates that biomass gasification could be deployed in the near-term timeframe.

Following the pattern of Ineos in England and Europe, to build its own private infrastructure of hydrogen fueling stations in England, the Chinese Institute of Geoscience and environment is working with large cities in China to build and infrastructure of hydrogen fueling stations in China. This Chinese zinstitue is the parent company of Grove Automotive so it is providing its own infrastructure to enable the sales of its own cars.

"The advantages and disadvantages of hydrogen fuel cells show us that if we can develop the necessary technologies to make this a widely available resource, then our future society could be very different from the one that we have today. It will give us more energy diversity while reducing the potential impact of carbon dioxide, methane, and other GHG emissions." (Miller, 2021) "increasing pressure—and—demand for zero-emission vehicles is likely to be the impetus for the hydrogen fuel cell revolution. Will the U.S. Lead or follow?" (Ohnsman, 2018)

BIBLIOGRAPHY

Accardi, M. 2022.Skip the batteries: Next-gen ram hd will run on hydrogen. Muscle Cars and Trucks.

Ahmed, H. 2021.Hydrogen; an eco-friendly energy source. Azonano. https://www.azonano.com/news.aspx?newsID=38288 retrieved 5-22-2024

Alaniz, A. 2021. Rumor claims mazda working on hydrogen-fueled rotary engine. Motor!. https://www.motor1.com/news/527241/mazda-developing-hydrogen-rotary-engine/ retrieved 9-5-2022

Ambrose, J. 2020. Airbus reveals plans for zero-emission aircraft fuelled by hydrogen. The Guardian. Theguardian.com/business/2020/sep/21/airb... retrieved 12-29-2020

Associated Prtess, 2022.GM expands market for hydrogen fuel cells beyond vehicles. Detroit Free Pree. https://www.freep.com/story/money/cars/general-motorss/2022/01/19/gm-hydrogen-fuel-cells-beyond-vehicles/6585832001/ retrieved7-2022

Associated Press. 2023. Honda's next hydrogen fuel cell vehicle to get stack developed with GM. Autoblog. https://www.autoblog.com/2023/02/02/honda-gm-hydrogen-fuel-cell-car/?guccounter=1&guce_referrer=aHROcHM6Ly93d3cuZ29vZ2xlLmNvbS88guce... Retrieved 5-10-2023

Arsenale. 2023. Peugeot 'no concept': a stunning hydrogen car concept. The Arsenale. https://thearsenale.com/blog/magazine/peugeot-no-concept-a-stunning-hydrogen-car-concept retrieved 5-16-2023

Autocar. 2018. Mercedes-Benz GLC F-Cell 2018 first ride. Autocar. Autocar. oo.uk/car-news/new-cars/mercedes-... retrieved 11-12-2020

Autocar.2018.Merceds committed to hydrogen fuel cell technology. Autocar. Autocar.oo.uk/car-news/motor-shows-paris-... retrieved 11-12-2020

AutomoStory. 2022. First hydrogen car. AutomoStory. https://www.automostory. com/first-hydrogen-car.htm retrieved 12, 2022

Automotive. 2022. Goodbye E-cars: BMWE wants to mass−produce hydrogen cars. Automotive. https://hydrogen-central.com-goodbye-e-cars-bmw-wants-mass-produce-hydrogen-cars/ retrieved 8-15-2022

Automotive, 2022.Out of electricity? This will be ford hydrogen giant.Automotive. https://hydrogen-central.com/out-electricity-ford-hydrogen-giant/ retrieved6-14-2023

Automotive.2023.Bosch engineering and ligier automotive present high-performance vehicle with a hydrogen engine at 24th race in lemans. Automotive.https://hydrogen-central.com/bosch-engineering-ligier-automotive-present-high-performance-vehicle-hydrogen-engine-at-24th-race—in-le-mans/ retrieved 6-14-2023

Automotive.2023.Hydrogen Toyota gazoo racing unveils "gr-h2 racing concept. Automotive. https://hydrogen-central.com/hydrogen-toyota-gazoo-racing-unveils-gr-h2-racing concept- le-mans-24-hours/ retrieved 6-14-2023

Autovista24. 2024. Namx plans hydrogen fuel-cell vehicle with swappable tanks. Autovista24. https://autovista24.autovistagroup.com/news/namx-plans-hydrogen-fuel-cell-vehicles-with-swappable-tanks/ retrieved 4-18-2024

Audi Media Center.2016. Hybrid fuel cell Audi h-tron. Audi Media Center. https://www.audi-mediacenter.com/en/hybrid-fuel-cell-slash-audi-h-tron-242 retrieved 9-1-2022

Ausick, P. 2020.10 Stocks riding the soaring demand for hydrogen fuel. Wall Street. 247Wallst.com/infrastructure/2020/07/24/10... retrieved 11-9-2020.

Autocar, co. UK. 2022. Nam huv is 542hp fcev with swappable hydrogen tanks. Fossil news. https://www.reddit.com/r/Real Tesla/comments/uo9v2k/namx_huv_is_542bhp_fcev_with_swappable_hydrogen/ retrieved 3-27-2024

Autocar Pro News. 2029.New hydrogen only car firm launches with pininfarina concept. autocarpro.in/news/-international/new-hydro... retrieved 12-21-2020

Auto power girl. 2007. 2007 ford fusion hydrogen 999 land speed record review and pictures. Auto-power-girl.com/cars-200-... retrieved 12-21-2020

Automotive. 2022. Out of Electricity? This will be Ford Hydrogen Giant. Automotive. https://hydrogen-central.com/out-electricity-ford-hydrogen-giant/ retrieven 12-29-2022

Autotrader.2022.Hydrogen-fueled Toyota mirai sets guiness world record-breaking 845-mile run. Autotrader. https://www.autotrader.com/car-news/hydrogen-fueled-toyota-mirai-sets-guiness-world-record-breaking-845-mie-run retrieved 7-2022h

Autotrader.2022.Land rover to build hydrogen fuel-cell defender. Autotrader. https://www.autotrader.com/car-news/d-rover-to-build-hydrogen-fuel-cell-defender retrieved 7-2022

Baker,D.R., Dezem, V., Freitas, Jr., G. 2021.Pipelind owners look to hydrogen as natural gas comes under attack. World Oil. Worldoil.com/news/2021/1/29/pipeline.... retrieved 3-16-2021

Baldwin, A. 2023. LeMans reaches 100 and looks to a hydrogen future. Reuters. https://www.reuters.com/sports/motor-sports-le-mans-reaches-100-looks-hyfrogen-future-2023-06-08/ retrieved 5-2, 2024

Baldwin, R. 2020.Hyundai's nexo is helping U.S. Expand hydroen fuel-cell r&d. Car and Driver. caranddriver.com/news/a30850029/hyundai-... retrieved 11-12-2020

Ballard. 2020. The future of clean transit is electric. Transit Bus. Ballard.com/market/transit-bus retrieved 11-15-2020

Ballard, 2021. Advantages of megawatt-scale power generation with fuel cells. Ballard's ClearGen-II system9https://info.ballard.com/talk-to-a-fuel-cell-retrieved 3-23-2021

Ballared. 2022. The future of clean transit is electric. Ballard Marketing. https://www.ballard.com/markets/transit-bus retrieved 8-18-2022

Ballard, 2023. Accelerating U.S. zero-emission fleet development through OCED funding. https://www.ballard.com/

blog-driving-zero-emission-fleet-deployments-through-oced-funding retrieved 9-22-2023

Ballard.2023.Stationary power generation, https://www.ballard.com/fuel-cell-solutions/fuel-cell-power-products/backup-power-systems retrieved 9-22-2023

Ballard.2023 Ballard announces orders from solaris for nearly 100 hydrogen fuel cell engines to power buses in Europe. https://https://www.ballard.com/aboout-ballard/newsroom/news-releases/2023/08/08/ballard-announces-orders-for-solaris-for-nearly-100-hydrogen-fuel-cell-e.... Retrieved 9-22-2023

Ballard. 2024. Hydrail: hydrogen fuel cell powered rail. Ballard. https://www.ballard.com/markets/rail retrieved 4-11-2024

Barlow, J. 2022. Alpine is developing a hydrogen fuel-cell racer for le mans. Top Gear. https://www.topgear.com/car-news/motorsport/alpine-developing-hydrogen-fuel-cell-racer-le-mans. Retrieved 7-24-2023

Barras, J. 2022. This hydrogen-powered racer is set to take on porsches and lambos, but what's the point. Electrek. Electrek.co/2022/01/26/ this hydrogen-powered-racer=is-set-to-take-on-prsches-and-lambos-but-whar's-the-point/ retrieved 7-3-2023

Bellis, M. 2019.Hydrogern fuel cells innovation for the 21st century. Thought Co. thoughtco.com/hydrogen-fuel-cells-19.... retrieved 3-28-2021

Berman, B. 2020. Daimler ends hydrogen car development because it's too costly. Electrek. Electrek.co/2020/04/22/daimler-ends-hydro... retrieved 12-18-2020

Biermann, R. 2023. Hyundai N version.74 to enter production as ppony coupe. Car Buzz. https://carbuzz.com/news/hyundai-n-vision-74-to-enter-production-as-pony-coupe retrieved 5-7-20

Biogradlija, A. 2022.Volkswagen throws the Diess, and its lands on hydrogen. Energy News. https://energynews.biz/volkswage-throws-the-diess-and-it-lands-on-hydrogen/ retrieved 9, 2022

Blanco, S. 2020. Hyperion XP-1: a hydrogen hypercar on a mission. Forbes. Retrieved 2-12-2023

Blanco, S. 2023.Honda reveals new details on 2024 cr-v powered by hydrogen. Car and Driver. https://www.caranddriver.com/news/a42796089/2024-honda-cr-v-powered-by-hydrogen-details/ retrieved 5-10-2023

Blain, L. 2022. Awesome-looking suv cleverly sidesteps hyrigen's biggest problem. Automotive. https://newatlas.com/automotive/namx-pininfarina-hydrogen-suv/ retrieved 3-27-2024

Bleier, E. 2016. Pininfarina's stunning hydrogen supercar concept is actually happening. Insidehook.com/article/vehicles/pininfarinas-... retrieved 12-21-2020

BMW Group. 2019. The BMW hydrogen next. Our fuel cell development vehicle. Bmwgroup.com/en/company/bmw-group-n... retirved 12-21-2020

BMW. 2022. BMW hydrogen fuel cell SUV to enter mass production as soon as 2025: executive. BMW. https://hydrogen-central-fuel-cell-suv-enter-mass-production-soon-2025-executiver retrieved 8015-2022

BMW. 2020. Hydrogen fuel cell cars: What you need to know. Bmw.com/en/innovation/how-hydrogen. retrieved 12-20-2020

BMW. 2020. innervation. Bmw.com/en/innervation/how-hydrogen-fuel... retrieved4-19-2021

BP news wires. Noria Corporation. 2022. Ford rolls out plug-in hybrid with hydrogen fuel cell. Reliable Plant. https://www.reliableplants.com/Read/4219/ford-rolls-out-plug-in-hybrid-wiTh-hydrogen-fuel-cell retrieved 12-31-2022

Bruce, C. 2022. Alpine A4810 concept imagines hydrogen-fueled supercar from 035. Motor 1. https://www.motor1.com/news/574447/alpine-a4810-hydrogen-hypercar-concept/ retrieved 3-27, 2024

Burgess, M. 2019. Rolls-Royce and Mercedes launch fuel cell project. H2-View. Chris.h2-view.com/story/rolls-rotce-and- mer... retrieved 11-12-2020

Burgess, M. 2020. Mercedes-Benz GenH2 truck revealed. H2-view.com/story/mercedes-benz-genh2Jr... retrieved 12-18-2020

California Fuel-cell Partnership. 2022. Fuel cell buses worldwide. California Fuel Cell Partnership. https://cafpcp.org/buses_trucks retrireved8-18-2022

Campbell, J. 2023. Is ford getting interested in hydrogen fuel vehicles? Hydrogen Fuel News. https://www.hydrogenfuelnews.com/ford-hydrogen-fuel-vehicles/8558650/ retrieved 6-14-2023

Capretto, A. 2023.Texas WILL be charging EV owners $200 for not paying gas taxes. Car Buzz. https://carbuzz.com/news/texas-will-charging-ev-owners-200-for-not-paying-gas-taxes retrieved5-7-2023

Carexpert/2021Audi writes off hydrogen for cars and SUVS. Car Expert. https://www.carexpert.com/au/car-news/audi-writes-off-hydrogen-for-cars-and-suvs retrieved 9-1-2022

Carey, N. 2021. German auto giants place their bets on hydrogen cars. Reuters. https:// www.reuters.com/technology/german-auto-giants-place-their-bets-hydrogen-cars-2022-09-22/ retrieved 9-1-2022

Carscoops, 2021.New opel and vauxhalkkl vicaro-e gets hydrogen powertrain with 249 mile range. Carscoops. https://www.carscoops.com/2021/05/new-opel-and vauxhall-vivro-e-gets-hydrogen-powertrain-with-249-mile-range/ retrieved 9-15-2023

Casey, T. 2023. Honda doubles down on hydrogen fuel cell electric car commitment. Clean Energy. https://cleantechnology.com/2023/02/05/honda-doubles-down-0on-hydrogen-fuel-cell-electric-car-commitment/ retrieved 5-10-2023

Catenacci, T. 2024. White House praised hertz for "accelerating the EV transition months before company began dumping EVs. Fox Business. https://www.foxbusiness.com/politics/white-house-praised-hertz-accelerating-ev-transition-months-company-began- dumping-evs retrieved 3-24-2024

Chapman, B. 2020.Hyundai and ineos announce deal to produce world's first hydrogen-powered suv. Independent. https://www.independent.co.uk/news/business/ hhhyundai-ineos-hydrogen-fuel-cell-car-grendaier-uk-b1760321.html retrieved 1-1-2024

Chatsko, 2020. Does the hydrogen economy have a pipeline problem?. Fool.com/ investing/2020/08/05/does-the-hy.... retrieved 3-16-2021

Chen, J. 2022. BMW and Toyota just gave hydrogen cars a big vote of conmfidence. INPUT. https://www.inputmag.com/tech/bmw-toyota-hydrogen-car-partnership-electric-vehicles retrieved 8-15-2022

Chen, Y. 2019. First grove hydrogen car produced in chongqing factory. Chongqing. Ichongqing.info/2019/04/14/first-grove-hydr... retrieved 12-21-2020

Cloete, S. 2020. Gas switching reforming: making hydrogen to balance variable wind, solar. Energy Post Events. Energypost.eu/gas-switching-reforming-mak... retrieved 11-26-2222

Cohen, A. 2020. Airbus unveils designs for hydrogen-powered aircraft which could be flying by 2035. Forbes. Forbes.com/sites/arielcohen/2020/09/23/airb... retrieved 12-29-2020

Collins, L. 2022. Exclusive hydrogen car company riversimple finds it has vastly exaggerated the range and performance of its FCEV. Retrieved 8-29-2023

Collins, L. 2023. Toyota pivots towards hydrogen trucks, admitting that its mirai fuel-cell car has not been successful. Accelerate Hydrogen. Retrieved 1-20-2024

Collins, L. 2023. Why has hydrogen car start-up namx ditched fuel cells for H2 internal combustion engines? Retrieved 3-27-2024

Collins, L. 2024. Hydrogen vehicle registrations fell by more then 50% in south korea in 2023, government figures reveal. Retrieved 4-18-2024

Collins, L. 2024. Exclusive hydrogen car sales in Japan have fallen by 83% over the past two years new figures show. Retrieved 4-18-2024

Corby, S. 2021. The history of hydrogen fuel cell cars. Carguide. https://www.carsguide.com.au/car-advice/the-hosory-fuel-cell-cars-85330 retrieved 12, 2022

Cummins Inc. 2021. Hydrogen fuel cell trains are on the fast track. Cummins Newsroom. https://www.cummins.com/news/2021/08/20/hydrogen-fuel-cell-trains-are-fast-track retrieved9-12-2022

Daly, M. Krisher, T. 2024.EPA issues new auto rules aimed at cutting carbon emissions, boosting electric vehicles and hybrids. Washington News. https://apnews.com/article/epa-electric-vehicles-emissions-limits-climate-biden-e6d581324af51294048df24269b5d20a retrieved 3-24-2024

Damian, A. 2006. The first hydrogen-powered sedan. Topspeed.com/cars/car-news/bmw-announc.... retrieved 12-21-2020

Davis, R. 2022. Toyota, Subaru and Mazda still betting on green combustion fuels amid EV shift. The Japantimes. Retrieved 5-17-2023

Daxue consulting. 2022. Hydrogen vehicles in china: will it overtake ev s? Daxue Consulting. https://daxueconsulting.com/hydrogen-vehicles-china/ retrieved 4-18-2024

Demarco blog. 2024. Hydrogenm. https://demarco-cryogenics.com/blog/liquid-hydrogen-storage/ retrieved 5-13-2024

Desjardins, L. 2020. Canadian railway to develop hydrogen-powered train. rcinet. ca/en/2020/12/19/canadian-railway-to... retrieved 1-1-2021.

Ding, D. and Xiao-Yu, W. 2024. Hydrogen fuel cell electric trains: technologies, current status, and future. Science Direct. https://www.sciencedirect.com/science/article/pii/S2666352X24000104 retrieved 4-11-2024

Discover, 2023. Arelectric cars worth the money? https://www.discover.com/personal-loans/resources/major-expenses/are-electric-cars-worth-it/#:-:text=All things considered%2C an EV,that's both fi... retrieved 11-10-2023

Dokso, A. 2024. The Japanese government has set an ambitious target of having 800,000 fuel cell hydrogen cars on the nation's roads by the end of the decade. H2 Energy News. https://energynews.biz/the-uphill-battle-for-hydrogen-cars-in-japan/#:-text=The Japanese government has set, in the past two years. Retrieved 4-128-2024

Dokso, A. 2024. Nissan ventures into hhydrogen mobility: unveiling venucia h2e fleet in china. Energy News. https://energynews.biz/nissan-ventures-intio-hydrogen-mobility-unveiling-venucia-h2e-fleet-in-china/ retrieved 4-22-2024

Domonoske, C. Copley, M. 2024. In a boost for EVs, EPA finalizes srict new limits on tailpipe emissions. Northern Public Radio. https://www.npr.orhg/2024/03/20/1239092833/biden-epa-auto-emissions-evs 3-24-2024

Dorgan, N & Ogg, M. 2022.Hydrogen car manufacturer H2X Global 'stepping on the gas' with eyes set on ipo. Business News. https://www.businesnewsaustralia.com/articles/hydrogen-car-manufacturer-h2x-global-steppiong-on-the-gas-with-eyes-et-on-ipo.html retrieved 6-2022

Dorofte, A. 2020. BMW group reaffirms commitment to hydrogen fuel cell technology. BMWBlog. Bmwblog.com/2020/03/30/bmw-group-reaff-... retrieved 12-21-2020

Dotzenrod, N, Fong, S., Kenny, V., Leung, M., Plaxco, J., Nathal, M., Saldana, S., Zuckerman, M., and Zuehlke, E. 2020. Natural Gas to Hydrogen (H). Northwestern. Processdesign.mccormick.northwestern.edu/... retrieved 11-26-2020

Dowling, J. 2019. Refueling stations. Caradvice.com.av/805257/hyundai-nexo-cert... retrieved 11-12-2020.

Drive Clean. 2021. Hydrogen Fuel Cell Electric Cars..https://driveclean.ca.gov/hydrogen-fuel-cell retrieved 5-5-23

Duff, M. 2022.Cloimbing behind the wheel of a hydrogen-powered fuel cell muscle car. Autoweek. https://www.autoweek.com/news/a41134652/hy-n-vision-74-hydrogen-fuel-cell-concept-drive/ retrieved 5-6-2023undai

Dunto. UK. 2023. Ford explores three-year hydrogen fuel cell e-transit trial that coud deliver increased range and uptime for operators. Ford Media Center. https://www.media.ford.com/content/fordmedia/feu/en/news/2023/05/10/ford retrieved 4-11-2023

Dyer, E. 2020. Hyperion XP-1 is a 221-mph hydrogen powered hypercar. Car and Driver. Aug. 12, 2020. https://www.caranddriver.com/news/a33572419/2022-hyperion-xp-1-hydrogen-hypercar-specs/ retrieved 2-12-2023

Eaton, K. 2009. Mercedes turns out fuel-cell B-class car, ready for public consumption. Fast Campany. Fastcompany.com/1343333/mercedes-turns-... retrieved 12-18-2020

Eddy, N. 2021. VW CEO says fuel cell cars not the answer to emissions-free mobility. Automotive News Europe. https://europe.autonews.com/automakers/vw-ceo-says-not-answer-emissions-free-mobility retrieved 9 2022

Edelstein, S. 2020. GM backs away from selling hydrogen powered passenger vehicles.Green Car Reports. Greencarreports.com/news/1128920_gm-bac... retrieved 12-15-2020

Edelstein, S. 2020. Mercedes-Benz F-cell SUV axed. Green Car Reports. Greencarreports.com/news/112791/_merced... retrieved 11-12-2020

Edelstein. S. 2020. Ballard reveals hydrogen fuel-cell tech developed with Audi. Green Car Reports. https://www.greencarreports.com/news/112624_ballard-reveals-hydrogen-fuel-cell-tech-developed-with-audi retrieved 9 1 2022

Edmunds. 2015.8 Things you need to know about hydrogen fuel-cell cars. Edmunds.com/fuel-economy/8things-you-need-to-know-about-hydrogen-fuel-cell-cars.html retrieved 3-16-2021

EERE. 2024. Why study hydrogen storage. https://www.energy/gov/eere/fuel cells/hydrogen-storage#:-text=Storage of hydrogen as a pressure is -252.8C rettieved 5-22-2024

EERE, 2024. Metal hydride storage materials. https://www.energy.gov/eere/fuel cells/metal-hydride-storage-materials retrieved 5-13-2024

EERE, 2023. Fuel cells. https://www.eneergy.gov/eere/fuelcells/fuel-cells retrieved 11-5-2023

EERE, 2023. Hydrogen Delivery. https://www.energy.gov/eere/fuelcells/ hydrogen-delivery retrieved 2-5-2023

EERE. 2022. Sustainable transportation. Retrieved 7-2022

EERE. 2021.Hydrogen pipelines. Energy.gov/eere/fuel/cells/hydrogen-pi.... retrieved 3-16-2021

EERE. 2021. 5 fast facts about hydrogen and fuel cells. Energy.gov/eere/articles/5-fast-facts-ab... retrieved 4-19-2021

EERE. 2020. hydrogen fuel basics. Energy.gov/eere/fuelcells/hydrogen.fu.... retrieved 4-19-2021

Eia. 2020. Hydrogen explained: uses of hydrogen. Eia.gov/energy explained/ hydrogen/use-of-h... retrieved 1-4- 2021

Eisenstein, P.A. 2021. GM enters the fuel cell business, will power Navistar trucks. Forbes Wheels. https://forbes.com/wheels/news/gm-enters-furl-cell-business-power-navistar-trucks/ retrieved 7-22-22

E-mobility, 2023. Challenge of batteries for heavy-duty evs. E-Mobility Engineering. https://www.emobility-engineering.com/challenge-of-batteries-for -heavy-duty-evs/ retrieved 11-10-2023

Environmental Technology. 2022, Hydrogen vehicle. https://www.mazda.com/en/ innovation/technology/env/hre/ retrieved 9-5-2022

Evans, D. Who invented the hydrogen fuel cell? Blog-post55.html retrieved 11-12-2020

Ewing, S. 2022. 2023 BMW iXr hydrogen first drive review: more than just a science project. CNET. https://www.cnet.com/roadshow/news/2023-bmw-ixr-hydrogen-first-drive-review/ retrieved 8-15-2022

Fast Tech. 2023. 8 Vehivle manufacturers working on hydrogen fuel cell cars. Fast Tech. https://www.fastechus.com/blog/vehicle-manufacturers-working-on-hydrogen-fuel-cell-vehicles retrieved 3-27-2024

FCHEA. 2023.Hydrogen as energy storage. https://www.fchea.org/hydrogen-as-storage retrieved 11-5-2023

FCHEA. 2022. Fuel cell basics. https://www.fchea.org/fuelcells retrieved 7-2022

FCHEA. 2022. Transportation. https://www.fchea.org/transportation retrieved 7-2022

Fender, K. 2020. Development of hydrogen-powered trains continues, but battery-powered equipment making more inroads. Trains Industry Newsletter. Trn. trsains.com/news-wire/2020/12/14... retrieved 1-1-2021

Fierro, V., Akdim, O., Mirodatos, C. 2003. O-board hydrogen production in a hybrid electric vehicle by bio-ethanol oxidative steam reforming over Ni and noble metal based catalysts. Creen Chem. Royal Society of chemistry. https://pubs.rsc.org/en/conternt/articlehtml/2003/gc/b208201m retrieved 8-14-2023

Foote, B. 2023. Hydrogen ford vehicles will remain niche products for now. Retrieved 7-26-2023

Ford, U.K. 2023. Ford announces three-year hydrogen fuel cell e-transit trial. Ford Media Center. https://media.ford.com/content/fordmedia/feu/gb/en/news/2023/05/09/ford-announces-three-year-hydrogen-fuel-cell-e-transit-trial.html retrieved 6-14-2023

Frangoul, A. 2020. Airbus announces concept designs for zero-emission, hydrogen-powered airplanes. Sustainable Energy. Cnbc.com/2020/09/21/airbus-announces-co... retrieved 12-29-2020

Frangoul, A. 2022.The most dumb thing: elon musk dismisses hydrogen as tool for energy storage. CNBC. https://www.cnbc.com/2022/05/12/tesla-ceo-elon-musk-dismisses-hydrogen-as=tool=for-energy-storage.html retrieved 9-2022

Frangoul, A. 2022. Renault reveals electric-hydrogen hybrid concept car, says it will have range of up to 497 miles. Sustainable Energy. https://www.cnbc.

com/2022/05/20/renault -says-electric-hydrogen-concept-will-have 497-mile-range.html retrieved9-2-2022

Frangoul, A. 2021. Jaguar Land Rover is developing a hydrogen-powered vehicle and plans to test it this year. Sustainable Future. https://www.cnbc.com/2021/06/15/jaguar-land-rover-is -developing-a-hydrogen-powered-vehicle-.html retrieved 9-12-2022

Frey, R. 2022. Riversimple rasa hydrogen. GLPAuto lgas. Info. Jttps://www.glpautogas.info/en/index84.html retrieved7-22-2022

Froehlich, P. 2007. Using hydrogen for gas chromatography. Laboratory Technology. . Labmanager.com/laboratory-technology/usin... retrieved 1-26-2021

Front Chem. 2022. Recent advances of metal borohydrides for hydrogen storage. Frontiers in Chemistry. https://www.ncbi.nim.nih.gov/pmcarticles/PMC9428915/ retrieved 5-16-2024

Froudokis, G.E. 2011. Hydrogen storage in nanotubes and nanostructures. Science Dighest. https://www.sciencedirect.com/science/article/pii/S1469702111701626 retrieved 5-22-2024

Gaton, B. 2021. The top ten electric vehicle myths that need to be debunked. The Driven. https://thedriven.io/2021/11/03/top-10-urban-myths-about-electric-vehicles-that-deserve-to-die/ retrieved 5-2023

Gardiner, G. 2023. Cryo-compressed hydrogen, the best solution for storage and refueling stations? Composites World. Retrieved 5-13-2024

Gilboy, J. 2020. BMW. Rretrieved 7-2020

Gitlin, J.M. 2020. General motors will engineer and build Nikola's hydrogen pickup. Arsdtechnica. Arstechnica.com/cars/2020/09/general-mota... retrieved 12-15-2020

Glon, R. 2021. Mazda is reportedly developing a hydrogen-burning rotary engine. Mazda News. https://www.autoblog.com/2021/08/16/mazda-hydrogen-burnong-rotary-engine/ retrieved 9-5-2022

Guardian, 2018. Germany launches world's first hydrogen-powered train. The Guardian. theguardian.com/environment/2018/sep/17/... retrieved 1-1-2021

Guardian. 2020.Ineos signs hydrogen fuel cell deal for its off-road vehicle. The Guardian. Theguardian.com/environment/2020/nov/23... retrieved 11-24-2020

Glon, R. 2020. Eveything you need to know about the nikola badger. Digital Trends. Digitaltrends.com/nikola-badger-releas... retrieved 11-12-2020

Glon, R. 2020. Hydrogen cars for sale. Digital Trends. Digitaltrends.com/cars/hydrogen-cars-f... retrieved 1-13-2021

Glpautogas.2023. What is green hydrogen and how is it procuced. Glpauto gas. https://www.glpautogas.info/en/how-hydrogen-cars-work.html retrieved 5-5-2023

Gluckman, D. 2007. Ford fusion hydrogen 999. Car and Driver. Caranddriver. com/news/a15152176/ford-fusi-... retrieved 12-21-2020.

GM Heritage. 2020. GM hydrogen fuel cell vehicles. Gmheritagecenter.com/featured/Fuel_cell_v... retrieved 12-15-2020

Golson, D. 2020. 2020 Huundai Nexo reviews: This hydrogen fuel-cell SUV deserves your attention. Road Show. Cnet.com/roadshow/reviews/2020-hyundai.... retrieved 1-13.2021

Grenadier, I. 2022.The five besty hydrogen cars to look forward to in Australia. Carsguide. https://www.carsguide.com.au/car-advice/the-five-best-hydrogen-cars-to-look-forward-to-in-australias-85333 retrieved6-2022

Green Car Congress. 2020. Honda receives type designation for level 3 automatic driving in Japan https://www.greencarcongress.com/2020/11/20201112-honda13.html retrieved 11-12-2020

Green Car Congress. 2020. BMW to pilot second-generation hydrogen fuel cell drives in small series from 2022. (https://www.grteencarcongress.com/2020/07/20200725-bmw.html retrieved 12-21-2020

Green Car Congress. 2021.Peugeot e-EXPERT hydrogen now in series production. Green Car Congress. https://www.greencarcongress.com/2021/05/2021527-peugeot.htms retrieved 9-2-2021

Green Car Congress. 2022. Renault Scenic Vision hydrogen hybrid concept offers 75% smaller carbon footprint than a conventional BEV. Green Car Congress.

https://www.grccncarcongress.com/2022/05/20220519-renault.html retrieved 9-2-2022

Green Car Congress2022. SoCalGas, Ford developing F-550 super duty hydrogen fuel cell electric truck.https://www.greencarcongress.com/2022/08/20220816-socalgas.html)

Green Car Congress. 2023. Ballard to suppy fuel cell systems to ford trucks for heavy-duty F-Max hydrogen prototype. Green Car Congress. https://www.grccncarcongress.com/2023/0808-ballard-html retrieved 4-11-2024

Green Car Reports. 2020Fuel Cell. Greencarreports.com/news/1099303_fiat-chr.... retrieved 3-28-2021

Grimmel, H. 2016. Nissan's fuel-cell solution: replace hydrogen with ethanol. Autoweek. Autoweek.com/news/green-cars/a1847996/n... retrieved 11-12-2020

Grove Hydrogen Automotive. 2020.About Grove. grove-auto.com/grovecompany retrieved 12-21-2020

Grove Hydrogen Automotive. 2020. Grove techology. grove-auto.com/technology retrieved 12-21-2020

Hall, K, 2022. Detroit Press, reprinted in "Ride," GM zeroes in on fuel cells. Chicago Tribune, May 29, 2022. Chicago, Il.

Halvorson, B. 2020. Racier -looking 2021 Toyota ready for more infrastructure. Green Car Reports. Greencarreports.com/news/1130322_racier-l.... retrieved 11-20-2020

Halvorson, B. 2021. Kia sustainability roadmap: hydrogen fuel-cell line up in 2028, leather and ICE phased out later. Green Car Reports. https://www.greencarreports.com/1134167_kia-sustainability-roadmap-hydrogen-fuel-cell-lineup-in-2028-leather-and-ice-phased-out-later retrieved 9-16=2022

Hanlon, M. 2004. BMW sets speed records with hydrogen fueled V12. Automotive. Newsatlas.com/bmw-sets-speed-records-with ... retrieved 12-21-2020

Harbaugh, J. 2016.We've got (rocket) chemistry, part 1. NASA. https://blogs.nasa.gov/Rockettology/tag/liquid-hydrogen/ retrieved 11-11, 2023

Hardigree, M. 2009.Jesse James sets hydrogen land speed record at 199.7 mph. Jalopnik. Jalopnik.com/jessie-james-sets-hydrogen-lan-... retrieved 12-21-2020

Hawkins, A. J. 2024. GM is developing a fleet of hydrogen-powered medium-duty trucks for DOE pilot. The Verge. https://www.theverge. com/2024/3/5/24090488/gm-hydrogen-fuel-cell-truck-fleet-pilot-doe retrieved 4-15-2024

Helmenstine, A.M. 2019. Deuterium Facts. Thought Co. thoughtco.com/facts-about-deuterium-6079.... retrieved 1-26-2021

Hill, J S. 2021.Australia's H2X to launch hydrogen fuel cell Warrego ute in November. The Driven. https://thedriven.io/2022/09/01/australias-h2x-to-launcj-hydrogen-fuel-cell-warrego-yte-in-november. Retrieve 6-2022

Hirschlag, A. 2020. Next stop, hydrogen-powered trains. Future Planet. Retrieved 12-31-2020

Holding, J. 2022. This alpine a4810 concept is powered by hydrogen. Topgear. https://www.topgear.com/car-news/future-tech/alpine-a4810-concept-powered-hydrogen retrieved 3-27-2024

Holding, J. 2023. Is toyota's hydrogen racing concept the future of le mans? Topgear. https://www.topgear.com/car-news/future-tech-/toyotas-hydrogen-racing-concept-future-;e-mans retrieved 6-14-2023

Honda news. 2023. Honda to begin U.S .production of fuel cell electric vehicle in 2024. Honda News. https://hondanews.com/en-US/releases/honda-to-begin-us-production-of-fuel-cell-electric-vehicles-in -2024 retrieved 5-10-2023

HT Auto Desk. 2020. Daimler mercedes-benz is stopping hydrogen car development. Here's why. HT Auto Desk. Auto.hindustantimes.com/auto/news/di... retrieved 12-18-2020

HT Auto. 2022. BMW and Toyota join forces to drive out new fuel cell cars: report. HT Auto. Retrieved 8-15-2022

Hurd, B. 2020. Hyundai's long-term strategy update includes fuel cell brand dubbed 'HTWO.' autoblog.com/2020/12/11/hyundai-2025-su... retrieved 12-16-2020

Hurlin, T. 2023. The 2025 24 hours of le mans will allow hydrogen race cars, Toyota delivers with the gr h2. Hemmings. https://www.hemmings.com/stories/toyota-gr-h2-concept/ retrieved 6-14-2023

Hydrogen Central. 2024. The hydrogen version of the Russian aurus has power reserve of 600Km and accelerates to 100Km/H in 4 s. Hydrogen Central Automotive. https://hydrogen-central.com/hydrogen-version-russian-aurus-power-reserve-600-km-accelerates-100-km-h-4-s/ retrieved 4-18-2024

Hydrogen concept cars. 2023. Fiat panda hydrogen concept. Hydrogen Cards Now. https://www.hydrogencarsnow.com/index.php/fiat-panda-hydrogen-concept/ retrieved 5-17-2023

Hydrogenfuelnews. 2023. Hydrogen fuel cell car unveiled as part of Chinese and Japanese automaker collab. World-Energy. https://www.world-energy.org/article/38528.html retrieved 4-22-2024

HydrogenHouse. 2022. Peugeot H2O fuel cell fire engine. Hydrogen House. https://hydrogenhouseproject.org/peugeot-h2o-fire-engine.html retrieved 5-16-2023

Hydrogenics. 2020. Hydrogenics. Hydrogenics.com/hydrogen-products-solutio... retrieved 11-15-2020

Hydrogenics. 2020. Fueling Solutions. Hydrogenics. Hydrogenics.com/hydrogen-product.... retrieved 11-21-2020

Hydrogen Europe. 2019. Hydrogen Buses. Hydroeurope.eu/hydrogen-buses retrieved 11-15-2020

Hydrogen Today. 2024. Nissan launches a fleet of hydrogen-powered vehicles in chhina. Hydrogen Today. https://hydrogentoday.info/en/home= retrieved 4-22-2024

Hydrotec. 2022. The zero emission, hydrogen fuel cell solution for land, air, and sea. Hydrotec. Retrieved 7-2021

Hydrotec. 2016. GM hydrogen fuel cells mark 50 years of development. Hydrotec. https//www.gmhydrotec.com/product/public/us/en/hydrotec/Home.detail.html retrieved 7-22-22

Hydroec. 2016. Mission-Ready Chevrolet Colorado ZH2 fuel cell vehicle breaks cover at U.S. Army Show. Hydrotec. https://www.gmhydrotec.com/product/public/us/en/hydrotec/Home.detailhtml. Retrieved 7-22-22

Ineos. 2020. Ineos aotomotive. Ineos.com/usinesses/ineos-automotive/ retrieved 11-24-2020.

Innovation. 2020. Hydrogen fuel cells, explained. Etrieved 7-2020

Irish Tech New. 2020. Ireland's first hydrogen fuel cell bus trial. Irish Tech News. Irishtechnews.ie/irelands-first-hydrogen-fuel... retrieved 11-15-2020

Ivanenko, A. 2020. It's time for Elon Musk to admit the significance of hydrogen fuel cells. Forbes. Forbes.com/sites/forbestechcouncil/2020/11.... retrieved 11-10-2020.

Jackson, E. 2020. Op-ed: Do hydrogen fuel cell buses make sense for cities? Streetblog. Usa.streetblog.org/2020/08/25/op-ed-do-h.... retrieved 11-15-2020

Jefferson, G.V. & Smith, J. 2022. HFCTA hydrogen fuel cell overview. Railway Age. https://www.railwayage.com/mechanical/hfcta-hydrogen-fuel-cell-overview/ retrieved 8-29-2023

Kable, G. 2022. VW involved in new hydrogen fuel-cell development. Wards Auto. https://www.wardsauto.com/industry-news-vw-involved-new-hydrogen-fuel-cell-development retrieved 5-6-2023

Kane, M. 2020. They hopeto achieve a commercially viable product with a range of up to 1,000 km (620miles) within 5-8 years. Motor1.com insideeevs.com/news/444480/mercedes-genh... retrieved 12-18-2020

Kane, M. 2023. Toyota u.s.hydrogen fuel ell car sales keep growing in q3 2023. Inside EVs. https://insideevs.com/news/693650/us-hydrogen-fuel-cell-sales-2023q3/ retrieved 1-24-2024

Kanttola, K. 2006. Ford Focus FCV. Hydrogencarsnow.com/index-php/f.... 11-15-2020

Kilbey, S. 2023. Ligier and Bosch reveal new hydrogen race car. Racer. https://racer.com/2023/06/08/ligier-and-bosch-reveal-new-hydrogen-race-car/ retrieved 6-15-2023

Kilbey, S. 2023. Le Mans delays hydrogen-powered class. Racer. https://racer.com/2023/11/09/le-mans-delays-hydrogen-ppowered-class/# retrieved 5-2-2024

Kim, S. 2023. The Korean Herald. https://www.koreaherald.com/view.php?ud=20230511000741 retrieved 4-18-2024

Kora, M. 2020. Are hydrogen fuel cell vehicles the future of auto? .ABC News. Abcnews.go.com/Business/hydrogen-fuel-cel... retrieved 12-15-2020

Koukouzas, N. 2017. Coal gasification. Science Direct. Sciencedirect.com/topics/engineering/coal-g... retrieved 11-28-2020.

Krishnakumar, T., Kiruthiga, A., Jozwiak, E., Moulaee, K. Neri, G. 2020. Development of ZnO-based sensors for fuel cell cars equipped with ethanol steam-reformer for on-board hydrogen production. Ceramics International. Science Direct. https://www.sciencedirect.com/science/article/abs/pii/S0272884220394770 retrieved 8-14-2023

La Grada. 2024. Ford has been investing in its hydrogen vehicles for years. Now its time has come. LaGrada. https://lagradaline.com/en/ford-hydrogen-vehivles/ retrieved 4-11-2024

Lambert, F. 2019. Hydrogen station explodes, toyota halts sales of fuel cell cars, is this the end? Electek. Electek.oo/2019/06/11/hydrogen-station-ex.... retrieved 11-21-2020

Lambert, F. 2020. Nikola (NKLA) admits to faking video of driving prototype in weak response to allegations. Electrek. Electrec.com/2020/09/14/nikola-nkla-admits-f... retrieved: 11-10-2020

Lambert, F. 2022. City cancels order of 50 hydrogen buses after realizing electric buses make so much more sense. Electrek https://electrek.co/2022/01/11/city-cancels-order-50-hydrogen-buses-after-realozing-electric-buses-best/ retrieved 8-8 2022

Lawson, A. 2021. Want long-term energy storage? Look to hydrogen. Power. https://www.powermag.com/want-long-term-energy-storage-look-to-hydrogen/ retrieved 11-5-2023

Lex, Kristan. 2022. Hydrogen fuel cells: Have hydrogen cars flopped? AutoTrader. https://www.autotrader.com.uk/content/news/hydrogen-cars-explained?refresh=true retrieved 12-27, 2022.

Lopez, J. 2020. GM hydrogen fuel cell vehicle on the horizon for the military. GM Authority. Gmauthority.com/blog/2020/10/gm-hydrog... retrieved 12-15-2020

Lykiardopoulou, I. 2022. Renault's ambitious concept car is hydrogen-electric and made from milk bottles. The next web. https://thenextweb.com/news/renault-scenic-concept-car-is-hydrogen-electric-made-from-milk-bottles retrieved 9-2-2022

Lyons, K. 2021. Jaguar Land Rover saysit's developing a hydrogen-powered prototype vehicle. The Verge. https://www.theverge.com/2021/6/16/22536905/jaguar-land-rover-hydrogen-powered-prototype-vehicle retrieved 9-12-2022

Macquarie. 2022. A clean start: south korea embraces its hydrogen future. Macquarie . https://www.macquarie.com/au/ev/insights/a-clean-start-south-korea-embraces-its-hydrogen-future.html retrieved 4-18-2024

Markus, F. 2020. Fertilizing fuel cells: One solution to the hydrogen infrastructure issue. Motor Trend. Motortrend.com/news/fuel-cell-fix-making-h.... retrieved 11-10-2020.

Matalucci, S. 2022. The hydrogen stream: Renault presents electric-hydrogen hybrid car with ranger of up to 800 km. pv-magazine. https://www,pv-magazine.com/2022/05/24/the-hydrogen-stream-renault-presents-ekectric-hydrogen-hybrid-car-with-range-0f-up-to-800-km/ retrieved 9-2-2022

MacKenzie, A. 2020. 2021 Ineos grenadier first look: is this land rover defender clone a rip-off or a reinvention? Motortrend. Motortrend.com/news/2021-ineos-grenadier... retrieved 11-24-2020.

MacKenzie, A. 2023. Ineos grenadier fuel cell ev suv first look: emission-free overlanding. Motor Trend. https://www.motortrend.com/news/ineos-grendadier-fuel-cell-ev-fcev-suv-first-look-review/ retrieved 5-17-2023

Martin, P. 2023. Nissan's Chinese joint venture launches first hyhdrogen fuel-cell car—but it will not be cheap. Dn Media Group. Retrieved 4-18-2024

Martin, P. 2024. Sales of hydrogen-powered vehicles in china rose bhy more than 70% in 2023. DN Media Group retrieved 4-18-2024

Martin, P. 2024. Hydrogen fuel-cell cars are "the next phase" after battery electric vehicles: Honda. DN Media Group. Retrieved 3-27-2024

Mazda, 2001.Mazda introduces new fuel cell electric vehicle, premacy fc-ev. Mazda Newsroom. https://newsroom.mazda.com/en/publicity/release/200102/0213e.html retrieved 9-2022

Mazda.2010.Mazda delivers premacy hydrogen re hybrid to Iwatani corporation for use in Kyushu. Mazda Newsroom. https://newsroom.mazda.com/en/publicity/release/2010/201001/100113a.html retrieved 9-2022

McEvoy, S. 2021. Do hydrogen powered racing cars hold the key to the future of clean mobility? Automotive World. Retrieved 7-24-2023

Mceachern, S. 2020. GM not planning hydrogen powered consumer vehicles for the time being. GM Authority. Gmauthority.com/blog/2020/07/gm-net-pla... retrieved 12-15-2020

Mercedes-Benz. 2020. F-cello: Behind the scenes of fuel cell mobility. Mercedes-benz.comen/innovation/electric/f... retrieved 12-18-2020

Mercedes-Benz. 2020. The new GLC F-cell. Mercedes-benz.com/en/vehicles/passenger-... retrieved 12-18-2020

Miller, K. 2020. Future of Working. Futureofworking.com/10advantages-and-di... retrieved 4-17-2021

Milner, P. 2021. Howto understand th ROI of your electric vehicle. Ultimarc blog. https://www.ultimarc.com/blog/return-on-investment-electric-vehicles/ retrieved 11-10-2023

Moon, M. 2021. Jaguar Land Rover to test a hydrogen fuel cell-powered Defender this year. Engadget. https://www.engadget.com/jaguar-land-rover-hydrogen-fuel-cell-defender-032948176.html retrieved 9-12-22

Moore, A. 2022. Volkswagen is developing a hydrogen car with a 1,250-mile range. Hyrdogen Fuel News. https://www.hydrogenfuelnews.com/volkswagon-hydrogen-car/8555838/ retrieved 5-6-2023

Motor Authority. 2024. Le Mans organizer reveals missionh24 hydrogen=electric racer. Motor Authority. https://www.motorauthority.com/news/1141263_le-mans-organizer-reveals-missionh24-hydrogen-electric-racer retrieved 5-2-2024

Motor Trend Staff. 2015. Ford delivers hydrogen focus fuel cell fleet. Motor Trend. Motortrend.com/news/news051028.... retrieved 11-15-2020

Morris, J. 2020. Why hydrogen will never be the future of electric cars., Forbes. Forbes.com/sites/james morris/2020 retrieved 11-15-2020

Mulach, J. 2021. Honda kills off the clarity due to low demand. Which Car. Ghttps://www.whichcar.com.au/car-news/honda-clarity-discontinued retrieved 5-10-2023

Muolo, D. and Thompson, C. 2017. Ford is pushing back its hydrogen car plans— here's why. Financeyahoo.com/news/ford-push... retrieved 11-15-2020

Nedelea, A. 2022. Peugeot launches E-expert hydrogen fuel cell van. Inside EVs. https://insideevs.com/news/597206/peugeot-e-expert-hydrogen-van-launched/ retrieved 5-16-2023

Newbold, J. 2023. How close is widespread adoption of hydrogen in motorsport. Motorsport. U.s.motorsport.com/lemans/news/how-close-is-widespread-adoption-of-hydrogen-in-mpotorsport/10447147/ retrieved 7-24-2023

Renault. 2021. New Peugeot e-EXPERT hydrogen, history: production has started. Renault Press Release. Retrieved 9-2-2022

New Foxy. 2019. Grove hydrogen automotive company and pininfarina announce wide ranging development partnership. New-foxy.com/2019/04/21/grove-hydrogen-a... retrieved 12-21-2020

Nikola Motors. 2020. Hydrogen advantagres. Nikolamotor.com/hydrogen retrieved 11-10-2020

Nils, A. 2022. Hydrogen fuel cell cars: everything you need to know. BMW.com. https://www.bmw.com/en/innovation/how-hydrogen-fuel-cell-cars-work.html retrieved8-15-2022

Nissan Motor Corporation. 2020. Next-generation fuel stack. Future Technology. Nissan-global.com/EN/technology/over... retrieved 11-12-2020

Nissan Motor Corporation. 2020. New X-trail Fuel cell vehicle. Nissan-global.com/EN/TECHNOLOGY/O... retrieved 1-13-2021

Noria Corporation, 2024. Ford rolls out plug-in hybrid with hydrogen fuel cell. Reliable Plant. https://www.reliableplant.com/Reed/4219/ford-rolls-out-plug-in-hybrid-with-hydrogen-fuel-cell retrieved 4-11-2024 retrieved 4-11-2024

Noriyuki, D. 2023. China's fuel cell vehicle push stalls out amid ev boom. Nikkei. https://asia.nikkei.com/Spotlight/Most-read-in-2023/China-s-fuel-cell-vehicle-push- stalls-out-amid-ev-boom retrieved 4-18-2024

Octa. 2024. Hydrogen fuel cell electric bus. Octa. https;??www.octa.net/about-octa/environmental-sustainability/fuel-cell/ retrieved 4-11-2024

Office of Energy Efficiency and Renewable Energy. 2020. Hydrogen Fuel Basics. Energy.gov/eere/fuelcells/hydrogen-fuel-bas... retrieved 11-24-2020.

Office of Energy Efficiency and Renewable Energy. 2020. Hydrogen production: biomass gasification. Energy.gov/eere/fuelcells/hydrogen-product... retrieved 11-27-2020

Office of Energy Efficiency and Renewable Energy. 2020. Hydrogen production: coal gasification. Energy.gov. Eere/fuelcells/hydrogen-product.... retrieved 11-28-2020

Office of Energy Efficiency and Renewable Energy. 2020.Hydrogen production: natural gas reforming. Energy.gov/eere/fuelcells/hydrogen-product... retrieved 11-26-2020

Office of Energy Efficiency and Renewable Energy. 2020. Hydrogen Fuel Basics. Energy.gov/eere/fuel/cells/hydrogen-fuel-bas... retrieved 1-4-2021

Office of Fossil Energy. 2020. Gasification systems. Energy.gov/fe/science-innovation/clean-coal... retrieved 11-28-2020

OhioState News, 2016. Ohio stste's all electric Venturi Buckeye Bullet 3 sets new landspeed record. News.osu.edu/ohio-state-all-electric-venturi... retrieved 3-25-2021

Ohnsman, A. 2019. Heavy-duty hydrogen: fuel cell trains and trucks power up for the 2020s. Forbes. forbes.com/sites/alanohnsman/2020/12/29/... retrieved 1-1-2021

Ohnsman, A. 2020. Nikola aims to close GM deal but has backup battery, fuel cell suppliers. Forbes. forbes.com/sites/alanohnsman/2020/11/09/... retrieved: 11-10-2020

Park, J. 2022. What fleets need to know about electric-truck batteries. HDT Truckinginfo. https;//www.truckinginfo.com/10166691/what-fleets-need-to-know-about-electric-truck-batteries retrieved 11-10-2023

Pappas, T. 2021. Kia to launch hydrogen-powered vehicles for the military before first passenger FCEV appears in 2028. Car Scoops. https://www.carscoops.com/2021/09/kia-to-launch-hydrogen-powered-vehicles-for-the-military-before-first-passenger-fcev-appears-in-2028/ retrieved 9-16-2022

Pappas, T. 2022. Hydrogen-powered hyperion XP-1 makes public debut with 2,000 HP and a 1,000-mile range. Car Scoops.com/2022/11/hydrogen-powered-hyperion-xp-1-makes-public-debut-with-2000-hp-and-a-1000-mile-range/ retrieved 2-12-2023

Parikh, S. 2023. 11 upcoming hydrogen fcevs with up to 500 miles of range. Top Electric SUV. https://topelectricsuv.com/featureed/upcoming-hydrogen-fcev-cars/ retrieved 5-17-2023

Patra, I. 2021. Jaguar Land Rover to develop hydrogen=powered concept car. The Hindu. https://www.thehindu.com/sci-tech/techn ology/jaguar-land-rover-to-develop-hydrogen-powered-concept-car/article34847180.ece retrieved 9-12-2022

Perry, T. 2018. Where does hydrogen for cars come from how's it produced? Green Car Future. Greencarfuture.com/hydrogen/hydrogen-fue... retrieved 11-24-2020.

Petrova, M. 2020. What's really going on at Nikola—an inside look at the truck maker in controversy. Pro. CNBC. Cnbc.com/2020/10/31/whats-happening-at-n... retrieved 11.10.2020

Peugeot. 2023. Hydrogen in production at Peugeot. Peugeot. Retrieved 5-16-2023

Pivovar, B. 20o20.Fuel Cells. NREL. Nrel.gov/hydrogen/fuel-cells.html retrieved 3-28-2021

Pleskot, K. 2015.Honda, Nissan, Toyota to cooperate on hydrogen infrastructure in Japan. Motor Trend. Msn.com//en-us/autos/other/honda-nissan-t... retrieved 1-13-2021

Quimby, T. 2022. Ford rolling out fuel cell F-550 utility truck. Hard Working Trucks. https://hardworkingtrucks.com/alternative-power/hydrogen-fuel-cell/article/15295981/ford-rolling-out-f550-utility-truck retrieved 12-31-2022

Rabenstein, G., Hacker, V. 2008. Hydrogen for fuel cells from ethanol by steam-reforming, partial-oxidation and combined auto-thermal reforming: a thermodynamic analysis. Journal of Power Sources. Science Direct. https://

www.sciencedirect.com/science/article/abs/pii/S0378775308015838 retrieved 8-14-2023

Ragonesi, O. 2007. HIL test systems for the bmw hydrogen 7. MicroNov AG. ni.com/en-us/innovations/care-studies/29/ni ... retrieved 12-21-2020

Rallypulse. 2024. The price of the luxury brand aurus powered by hydrogen is clear. Rallypulse—Rally and Auto News. https://rallypulse.com/news/the-price-of-the-luxury-brand-piwered- by hydrogen-is clear.html retrieved 4-20-2024

Ramey, J. 2020. Will the ineos grendier get hydrogen fuel cell technology from hyundai? Autoweek. Autoweek.com/news/green-cars/a34771969/... retrieved 11-24-2020

Ramey, J. 2020. BMW will offer hydrogen fuel-cell X5 suv in 2022. Autoweek. Hearst Auto, Inc.

Randall, C. 2021.Production start for the fuel cell opel vivaro. Electrive.com. https://www.electrive.com/2021/12/09/prduction-start-for-the-fuel-cell-opel-vivaro/ retrived5-17-2023

Randall, C. 2020. Daimler plans H2 truck with 1,000 km range. Electrive.com. Electric.com/story/2020/09/16/daimler-reveals-pl... retrieved 12-18-2020

Randall, C. 2022. Hyperion motors presents 1.5 mw fuel cell hypercar. Electrive. com. https://www.electrive.com/2022/11/23/hyperion-motors-presents-1-5-mw-fuel-cell-hypercar/ retrieved 8-22-2023

Ranga, 2015. 10 Use sof hydrogen: for industry and everyday life. Study read. Studyread.com/uses-of-hydrogen/ retrieved 1-4-2021

Railway technology.2021. Russian railways and partners to develop hydrogen fuel cell locomotives. Railway Technology. https://www.railway-technology.com/neews/russian-railways-partners-locomotive/?cf-view retrieced 4-18-2024

Reyes, A. 2022. The futuristic hydrogen-powered ford that has almost been forgotten. Slashgear https://www.slashgear.com/927725/the-futuristic-hydrogen-powered-ford-that-has-almost-been-forgotten/ retrieved 12-31-2022

Ridgebackpilot. 2024.Are hydrogen-powered cars really an alternative to evs? Mache Forum. https://www.macheforum.com/site/threads/

are-hydrogen-powered-cars-really-an-alternative-to-evs.33599/ retrieved 4-11-2024

Rivard, E, Trudeau, M, Zaghib, K. 2019. Hydrogen storage for mobility: a review. MDPI. https://www.ncbi.nim.nih.gov/pmc/articles/PMC6630991/ retrieved 5-7-2024

Riversimple. 2022. Riversimple rasa lightweight hydrogen car revived with siemens partnership. Green Car. https://www.greencarreports.com/news/1131296_riversimple-rasa-lightweight-hydrogen-car-nudged -toward-production-with siemens-partnership retrieved 8-29-2023

RPnews wire.2022. Ford rolls out plug-in hybrid with hydrogen fuel cell. Reliable Plant. https://www.reliableplant.com/Read/4219/ord-rolls-out-plug-in hydrogen-fuel-cell retrieved 7, 2022

Ryan, C. 2020. Airbus unveils hydrogen designs for zero-emission flight. Bloomberg News. Bloomberg.com/news/articles/2020-09-21/ai... retrieved 12-29-2020

Sara. 2023. Toyota unveils hydrogen racing concept for le mans. Race Tech. https://www.racetechmag.com/2023/06/toyota-unveils-hydrogen-racing-concept-for-le-mans/ retrieved 6-14-2023

Sergeev, A. 2022. Ford patents hydrogen-fueled combustion engine.ford-hydrogen-combustion-engine-patent/ retrieved 7-2022 Motor 1. https://www.motor1.com/news/574337/

Sergeev, A. 2022. Toyota, Subaru among six companies researching environmentally friendlyfuerls. Motor 1. https://www.motor1.com/news/599625/toyota-subaru-working-green-fuels/ retrieved 5-17-2023

Sergeev, A. 2023.Ineos grenadier hydrogen pushed back due to lack of infrastructure. Motor 1. https://www.motor1.com/news/633044/ineos-grenadier-hydrogen-cancelled/ retrieved 1-24-2023

Sheldon, A. 2022.Are electric car batteries bad for the environment? https://magazine.northeast.aaa.com/daily/llife/cars=trucks/electric-vehicles/are-electric-car-batteries-bad-for-the-environment/ retrieved 9-2022

Sherer, K. 2007. Hydrogen powered ford fusion 999 tops 207 miles per hour. Automotive. Newatlas.com/hydrogen-powered-ford-fusion.... retrieved 12-21-2020

Shet, S.P., Priva, S.S., Sudhakar, K., &Tahhir, M. 2021. A review on current trends in potential use of metal-organic framework for hydrogen storage. Science Digest. https://www.sciencedirect.com/science/article/abs/pii/SO360319921000331 retrieved 5-23-2024

Sikarwar, V. & Zhao, M. 2017. Biomass gasification. Earth systems and Environmental Sciences. Researchgate.net/publication/315849347_Bio... retrieved 11-27-2020

Smartcharger. 2020. Smartcharger. speedace.ifo/education/Smar... retrieved 12-21-2020

Smirnow, A. 2022. Report: Electric Ram heavy-duty truck with a hydrogen fuel-cell option is rumored. The fast lane truck. https://tfltruck.com/2022/04/report-electric-ram-heavy-duty-truck-with-a-hydrogen-fuel-cell-option-is rumored/ retrieved 5-6-2023

Smirnow, A. 2023. This hydrogen fuel cell may power your future ram HD truck or large van-stellantis electrification plan. TFT Truck. https://tfttruck.com/2023/01/ths-hydrogen-fuel-cell-may-power-youir-future-ram-hd-truck-or-large-van-stellantis-electrificationm-plan/ retrieved 5-16-2023

Smith, B. 2023. Hydrogen-powered alfa Romeo p7 hypercar concept needs to become a reality. Techeblog. https://www.techeblog.com/hydrogen-powered-alfa-romeo-p7-hypercar-concept/ retrieved9-15-2023

Snytnikov, P.V., Badmaev, S.D., Volkova. G.G., Ptemkin, D.I., Zryyanova, M.M., Belyaev, V.D. 2012. Catalysts for hydrogen production in a multifuel processor by methanol, dimethyl ether, and bioethanol steam reforming for fuel cell applications. International Journal of Hydrogen Energy. Science Direct. https://www.sciencedirect.com/science/article/abs/pii/S0360319912004806 retrieved 8-14-2023

Southern California Gas Company. 2022. SoCalGas joins Ford to reduce emissions with cutting edge Ford F-550 super duty hydrogen fuel cell electric truck. SoCalGas. Retrieved 12-31-2022

Spowers, F. 2022.Riversimple rasa unveiled. Riversimple. Retrieved 8-29-2023

Stellantis, 2023. It's all about hydrogen fuel cells. The Auto Channel. https://www.theautochannel.comnews/2023/05/17/1285803-it-s-all-about-hydrogen-fuel-cells-stellantis-to-acquire.html retrieved 5-17-2023

Stellantis. 2021. Into the hydrogen-based future now., with opel vivaro-e hydrogen. Retrieved 9-15-2023

Stellantis, 2021. Hydrogen Fuel Cell Technology. Stelanttis.com/en/technology/ hydrogen-fuel-cell technology retrieved 5-5-3023

Stevens, M. 2015. Maserati boss rules out electric models: hybrids coming, but hydrogen the future. Drive. https://www.drive.com.au/news/maserati-boss-rules-out-electric-models-hybrids-coming-but-hydrogen-the-future/ retrieved 5-18-2023

Stopford, W. 2022. BMW boss says hydrogen vehivles will be the next trend. Car Expert. https://www.carexpert.com.au/car-news/bmw-boss-says-hydrogen-vehivles-will-be-the-next-trend retrieved8-24-2023

Szymkowski, S. 2019. 2020 Honda clarity fuel cell will work better in the cold now. Road Show. Chet.com/roadshow/news/2020-honda-clarit... retrieved 11-12-2020.

Tariq, A. 2024. 15 Hydrogen cars to look out for. Top Speed. https://www. topspeed.com/hydrogen-cars-to -look-out-for retrieved 3-27-2024

Tengler, S. 2022. New study suggests ongoing issue with selling electric vehicles. Forbes. https://www.forbes.com/sites/stevetengler/2022/01/11/new-study-suggests-ongooing-issue-with-selling-electric-vehicles?sh+4f226b983540 retrieved 11-10-2023

Taylor, A. 2016. Riversimple rasa review: is this hydrogen car the future—or just a gimmick? Ars Technica. https://arstechnica.com/cars/2016/04/riversimple-rasa-hydrogen-car-review/ retrieved 8-29-2023

Telematics. 2020. Shell, toyota, and honda plan expansion of hydrogen refueling network in california. Telemasticsnews.info/2020/11/10/shell-toyota... retrieved 11-12-2020

Thompson, A. 2020.Where are all the hydrogen cars we were promised? Popular Mechanics. Retrieved 11-10-2020

Thought Co. 2020. 10 Interesting facts about radioactive tritium. Thought Co. thoughtco.com/facts-about-tritium-60/915 retrieved 1-26-2021

Timperley, J. 2021. When it comes to buses, will hydrogen or electric win. Science. https://www.wired.com/story/future-buses-hydrogen-electric/ retrieved 8-18-2022

Tingwall, E. 2024. The next diesel? GM and honda start u.s. production of hydrogen fuel cells. Motor Trend. https://www.motoretrend.comnews/honda-general-motors-hydrogen-fuel-cell-production-start/ retrieved 4-15-2024

Toyota.com/Mirai/stations.html retrieved 11-21-2020

Truett, R. 2001. Chrysler unveils zero-emission, fuel cell powered minivan. Autoweek. Autoweek.com/news/a2129571/chrysler-unv...

TWI. 2022. What is a hydrogen train and how do they work? TWI-Global. https://www.twi-global.com/technical-knowledge/what-is-a-hydrogen-train retrieved 9-12-2022

TWI.2022. What are the pros and cons of hydrogen fuel cells?. TWI-Global. https://www.twi-global.com/technical-knowledge/faqs/what-are-the-pros-and-cons-of-hydrogen-fuel-cells. Retrieved 9-18-2022

U.S. Department of Energy. 2020. Hydrogen fueling infrastructure Development. Afdc.energy.gov/fuels/hydrogen_infrast... retrieved 11-14-2020.

U.S. Department of Energy. 2020. Hydrogen fueling stations. Afdc.energy.gov/fuels/hydrogen_stations.html retrieved 11-21-2020

Usesof.2020. Uses of Hydrogen. Usesof.net/uses-of-hydrogen.h... retrieved 1-4-2021

University of Copenhagen. 2020. Fuel cells for hydrogen vehicles are becoming longer lasting. Fuel cells for hydrogen vehicles are becoming longer lasting. (2020, August 24) retrieved 9 November from https://phys.org/news/2020-08-fuel-cells-hydrogen-vehicles-longer.html

Uwaoma, P. 2022. The real reason why honda clarity failed. Hot Cars. https://www.hotcars.com/the-real-reason-why-honda-clarity-faied retrieved 5-10-2023 electricity ultimately makes battery-electric cars less green than fuel-cell vehicles." (Greencars,2020)

Vaidya, P.D., Rodrigues, A.E., Insight into steam reforming of ethanol to produce hydrogen for fuel cells. Chemical Engineering Journal. Science Direct. Htps://

www.sciencedirect.com/science/article/abs/pii/S138594705004808 retrieved 8-14-2023

Valdes-Dapena, P. 2020. This hydrogen-powered supercar can drive 1,000 miles on a single tank. CNN Business. https://www.cnn.com/2020/08/12/success/hyperion-xp1-hydrogen-powered-supercar/index.html retrieved 5-16-2023

Vaughn, M. 2020. 10 Things you didn't know about hydrogen fuel cell vehicles in America. Aotoweek. Hearst Autos, Inc.

Vaughn, M. 2009. Jesse James brings hydrogen land-speed record back to America. Autoweek. Autoweek.com/news/a2018701/jesse-james-... retrieved 12-21-2020

Viessmann.2020. How efficient are hydrogen fuel cells? Viesmann.co.uk/heating-advice//how-efficie... retrieved 11-24-2020

Voelcker, J. 2022. Hydrogen fuel-cell vehicles: everything you need to know. Car and driver. https://www.caranddriver.com/features/a41103863/hydrogen-cars-fcev/ retrieved 5-7-2023

Voelcker, J. 2020. Whywe still can't deliver on the promise of hydrogen cars. The Drive. Retrieved 7-2020

Voelcker, J. 2016. Nissan takes a different approach to fuel cells : ethanol. Greencarsreport.com/ news/1104467_nissan-t.... retrieved 1-13-2021

Volkswagen Group. 2019. Hydrogen or battery? A clear case, until further notice. Volkswagen. https://www.volkswagenag.com/news/stories/2019/08/hydrogen-or-battery-that -is-the-question.html/ retrieved 5-6-2023

Walker, K. 2020.Where are hydrogen fuel cells used. Azocleantech.com/article.asapxArticleID+333 retrieved 4-19-2021

Wallace, F. 2023. Unleashing the power: Liquid hydrogen fuels debut race car. Hydrogen Fuel News. https://www.hydrogenfuelnews.com/liquid-hydrogen-first-race-car/8558968/ retrieved 7-24-2023

Wand, G. 2017. Ford believes in hydrogen and fuel cell future. Hydrogen Cars Now. Hydrogencarsnow.co... retrieved 11-15-2020

Watlins, G. 2023. Le mans wants hydrogen-only top class by 2030. Motorsport. https://us.motorsport.com/le-mans-wants-hydrogen-only-top-class-by-2030/10474312/ retrieved 5-2-2024

Wayland, M. 2024. GM, honda begin u.s. fuel cell production in step toward replacing diesel. CNBC Auto. https://www.cnbc.com/2024/01/25/gm-honda-begin-us-fuel-cell-production-html retrieved 4-15-2024

WBUR. 2017. Why GM and Honda are betting millions on hydrogen-powered cars. WBUR Here & Now. Wbur.org/hereandnow/2017/03/03/general-... retrieved 12-15-2020

West, L. 2019. Pros and cons of hydrogen fuel-cell vehicles. Treehugger.

WHA.2023.10 Hydrogen fuel cell applications you might not know. https://wha-international.com/10-hydrogen-fuel-cell-applicaions-ypou-might-not-know/ retrieved 2-5-2023

Wiles, R. 2020. After resignation of founder, what's next for Nikola's truck-manufacturing plans in Arizona? Arizona Republic. Azcentral. Azcentral. com/story/money/business/2020/O.... retrieved: 11-10-2020

Williams, B. 2022. Hyperion XP-1 hydrogen car unveiled with 1,000-mile range. Hydrogen Fuel News. https://www.hydrogenfuelnews.com/hydrogen-car-hyperion-xp-1/8556084/ retrieved 8-23-2023

Willighe, A. 2022. 4 ways of storing hydrogen from renewable energy. Spectra. https://spectra.mhl.com/4-ways-of-storing-hydrogen-from-renewable-energy retrieved 5-13-2024

Winton, N. 2020. Could hydrogen fuel cells revive, threaten battery technology in cars? Forbes. Forbes.com retrieved 11-10-2020

Youd, F. 2021. Next stop, hydrogen? The future of train fuels. Railway-technology. https://www.railway-technology.com/analysis/next-stop-hydrogen-the-future-of-train-fuelds/ retrieved 9-12-2022

YPTE, 2022.Electric Cars. YPTE. https://ypte.org.uk/factsheets/electric-cars/ what-are-the-downsides-to-electric-cars retrieved 9-2022c

Zachariah, B. 2021. Alfa Romeo to electrify, as billion-dollar Giorgio architecture abandoned-report. Drive. https://www.drive.com.au/news/alfa-romeo-to-electrify-as-billion-dollar-architecture-abandoned-report/ retrieved 5-6-2023

Zelenak, V. & Ivan, S. 2021. Factors affecting hydrogen adsorptions in metal-organic frameworks: a short review. NCBI, https://www.ncbi.nlm.nih.gov/ pmc/articles/PMC8303527/ retrieved 5-23-2024

Zhang, B., Sun, Y., Xu, H., & He, X. 2023. Hydrogen storage mechanism of metal-organic framework materials based on metal centers and organic ligands. Wiley online. https://onlinelibrary.wiley.com/doiu/full/10.1002/cnl2.91 retrieved 5-23-2024

Zukowski, D. 2024. Hertz reverses course on electric vehicles. Utility Dive. https://www.utilitydive.com/news/hertz-sells-electric-vehicles-replace-wih-gas-engine/ retrieved 3-24-2024

Zetsche, D. 2020.Mercedes deniesCEO said mercedes said they will turn away fromhydrogen fuel vehicles (update). Greencarreports.com/news/1109713_did-me... retrieved 12-18-2020

Zino, K. 2009. Hydrogen fueled land speed runbeats BMW?. Thedetroitbureau.com/2009/06/hydrogen-fu.... retrieved 12-21-2020

ABOUT the AUTHOR

DR. BOB KAPHEIM RECENTLY PUBLISHED
"The Lithium Rush is On This book describes the lithium-ion battery, its uses, advantages and disadvantages and the coming demand for
lithium for electric cars which has caused a worldwide search to find and
mine lithium. During his career, Dr. Kapheim has served as an educator,
graduate instructor, presenter, consultant, and author. Kapheim currently is
a professor of biology at Judson University and formerly a faculty member
in the School of Education at Saint Xavier University in Chicago, Illinois.
As an award-winning educator, he taught biology and general science at
the high school level. During his tenure as a secondary school teacher, he
won numerous awards including teacher of the year in his district, three
national awards for innovative biology lessons, and was part of a delegation
representing the United States in an exchange with Russia and Poland. He
has been a national presenter at science and gifted conferences. In addition,
he has presented at state level science and gifted conferences. Formerly
he served as the Head of the Upper School and Dean at the Science and
Arts Academy, a private school for gifted. While at the Science and Arts
Academy he designed and installed the "Interim" a highly innovative curriculum for gifted children.

Bob began teaching graduate courses for Saint Xavier University through IRI Skylight. This experience led to his becoming the Director of Curriculum and Instruction for Pearson Education as he has managed field-based cohort Master's degree programs, discrete graduate level courses for Saint Xavier University, and distance education programs for both Saint Xavier and Drake University. Following his position at Pearson Education Dr. Kapheim became the Dean of Math and Science at DeVry University in Addison, Illinois. As a consultant for IRI Skylight he has given programs throughout the United States on brain-based learning, multiple intelligence, cooperative education, authentic assessment, positive discipline, 20th Century education, instrumental enrichment, and Socratic dialogue. Noted for his innovative curriculum designs, Dr. Kapheim has authored the book "Question the Thought," which provides teachers with strategies for effective classroom discussion through Socratic dialogue. Cricket Creek a book of essays on ecology and REfS, Reading Exercises for Science a book designed to help struggling readers in science.

Contact author at
bobkapheim@yahoo.com